NUREG/CR-7135

Compensatory and Alternative Regulatory MEasures for Nuclear Power Plant FIRE Protection (CARMEN-FIRE)

Draft Report for Comment

Office of Nuclear Regulatory Research

AVAILABILITY OF REFERENCE MATERIALS
IN NRC PUBLICATIONS

United States Nuclear Regulatory Commission

Protecting People and the Environment

NUREG/CR-7135

Compensatory and Alternative Regulatory MEasures for Nuclear Power Plant FIRE Protection (CARMEN-FIRE)

Draft Report for Comment

Draft Manuscript Completed: February 2013
Date Published: June 2013

Prepared by:
K. Sullivan
Brookhaven National Laboratory
Nuclear Science and Technology Department
P.O. Box 5000
Upton, NY 11973

F. Gonzalez, NRC Project Manager

NRC Job N6760

Office of Nuclear Regulatory Research

COMMENTS ON DRAFT REPORT

Any interested party may submit comments on this report for consideration by the NRC staff. Comments may be accompanied by additional relevant information or supporting data. Please specify the report number **NUREG/CR-7135** in your comments, and send them by the end of the comment period specified in the *Federal Register* notice announcing the availability of this report.

Addresses: You may submit comments by any one of the following methods. Please include Docket ID **NRC-2013-0103** in the subject line of your comments. Comments submitted in writing or in electronic form will be posted on the NRC website and on the Federal rulemaking website http://www.regulations.gov.

Federal Rulemaking Website: Go to http://www.regulations.gov and search for documents filed under Docket ID **NRC-2013-0103**. Address questions about NRC dockets to Carol Gallagher at 301-492-3668 or by e-mail at Carol.Gallagher@nrc.gov.

Mail comments to: Cindy Bladey, Chief, Rules, Announcements, and Directives Branch (RADB), Division of Administrative Services, Office of Administration, Mail Stop: TWB-05-B01M, U.S. Nuclear Regulatory Commission, Washington, DC 20555-0001. Faxes may be sent to RADB at 301-492-3446.

For any questions about the material in this report, please contact: Felix Gonzalez, Reliability and Risk Engineer, and 301-251-7596 or by e-mail at felix.gonzalez@nrc.gov.

Please be aware that any comments that you submit to the NRC will be considered a public record and entered into the Agencywide Documents Access and Management System (ADAMS). Do not provide information you would not want to be publicly available.

ABSTRACT

Each commercial nuclear power plant operating in the United States has a comprehensive Fire Protection Program (FPP) that has been reviewed and approved by the U.S. Nuclear Regulatory Commission (NRC). [1] The purpose of the FPP is to prevent the occurrence of fire and minimize radioactive releases to the environment in the unlikely event a significant fire were to occur. To achieve these objectives, the FPP integrates a number of plant designs and operating features (i.e., systems, components, personnel and administrative controls) needed to provide defense-in-depth protection of the public health and safety. This means that the FPP includes measures that are directed at reducing the likelihood of fires and explosions; rapidly detecting and suppressing those fires that may occur, and ensuring the capability to achieve and maintain safe shutdown conditions in the event of a significant fire.

Over the life of a plant, certain elements of the FPP may be found to be incapable of performing their intended function (i.e., become impaired). Ensuring appropriate actions are promptly taken to mitigate the effect of such impairments until permanent corrective actions can be completed is a key element of the FPP. Such actions are called "compensatory measures."

For most impairments, appropriate measures are specified in the approved FPP. However, since it is not possible to address all conceivable plant conditions, the actions specified in the FPP may not be appropriate in all cases. In such instances, an alternate compensatory measure is identified by the licensee on a case-by-case basis. The selection of a specific type of compensatory measure for a given impairment should be based on its ability to offset the degradation in defense-in-depth created by the inoperable, degraded, or nonconforming condition. Thus, appropriate alternate measure(s) may consist of wide range of options; from enhancing the measure specified in the FPP (e.g., placing additional controls over combustible materials) to the use of technologically advanced fire protection systems that were not available at the time the FPP was approved.

Employing compensatory measures, on a short-term basis, is an integral part of NRC-approved fire protection programs. However, compensatory measures are not expected to be in place for an extended period of time. The NRC staff expects that the corrective action(s) will be completed, and reliance on the compensatory measure eliminated, at the first available opportunity, typically the first refueling outage. Thus, a compensatory measure that is in place beyond the next refueling outage (typically 18 – 24 months) is considered to be a "*long-term compensatory measure.*"

This report is intended to serve as a reference guide for agency staff responsible for evaluating the acceptability of interim compensatory measures provided to offset the degradation in fire safety caused by impaired fire protection features at nuclear power plants. The report documents the history of compensatory measures and details the regulatory framework established by NRC to ensure they are appropriately implemented and maintained. This report

---

[1] As defined in Generic Letter 81-12, an approved Fire Protection Program includes the fire protection and post-fire safe shutdown systems necessary to satisfy NRC guidelines and requirements; administrative and technical controls; the fire brigade and fire protection related technical staff; and other related plant features which have been described by the licensee in the FSAR, fire hazards analysis, responses to staff requests for additional information, comparisons of plant designs to applicable NRC fire protection guidelines and requirements, and descriptions of the methodology for assuring safe plant shutdown following a fire.

also explores technologies that did not exist when the current plants were licensed such as video-based detection, temporary penetration seals and portable suppression systems which under certain conditions may provide an effective alternative to traditional measures specified in a plant's approved fire protection program.

TABLE OF CONTENTS

LIST OF FIGURES

LIST OF TABLES

Table	**Page**

EXECUTIVE SUMMARY

During the initial development of the U.S. commercial nuclear power program, the level of fire protection adopted by a nuclear power plant (NPP) was generally found to be acceptable if it complied with applicable national and local fire codes, as evidenced by an acceptable rating from its fire insurance underwriter. Consequently, the fire protection features at NPPs were very similar to those of conventional, fossil-fueled, power generating stations. Fire protection continued to be evaluated on this basis until a major fire at the Browns Ferry Nuclear Power Plant on March 22, 1975 prompted significant changes in the NRC's fire-safety regulatory framework.

A key recommendation of a Special Review Group (SRG) tasked by the Commission to investigate the Browns Ferry fire, was the need for each operating plant to establish a comprehensive fire protection program (FPP) that integrates the multiple levels of protection established by the time-honored safety concept known as *"defense-in-depth."* The incorporation of this recommendation into the NRC's regulatory framework required each operating plant to establish a FPP that integrates the structures, systems, components, procedures, and personnel needed to: (1) Prevent fires from starting; (2) Detect rapidly, control, and extinguish those fires that do occur; and, (3) Ensure that fire will not prevent the performance of necessary safe shutdown functions, and will not significantly increase the risk of radioactive releases to the environment. The aim is to achieve a suitable balance between each layer of protection. Increased strength, redundancy, performance, or reliability of one layer of protection can compensate in some measure for deficiencies in the others.

Consistent with the defense-in-depth safety concept, the SRG also recommended that an adequate level of fire safety be maintained when a fire protection feature is degraded or disabled. The Commission addressed this recommendation by allowing licensees, through their approved fire protection program[2], to implement temporary corrective actions called *"compensatory measures"* or *"interim compensatory measures"* until the functionality of degraded or disabled equipment is restored. The objective of these measures is to temporarily offset the reduction in fire safety caused by the impairment.

The incorporation of SRG recommendations into the NRC's regulatory framework resulted in a comprehensive FPP at each plant that integrates design features needed to provide an appropriate balance between each element of defense-in-depth and administrative controls (e.g., policies and procedures) to govern all fire protection activities, including the establishment of interim compensatory measures that are appropriate for the specific hazards, compatible with plant operations, and are properly implemented and maintained.

To assure that fire protection systems and features are available and in proper working condition, the Commission initially imposed Technical Specifications (TS) for certain fire

---

[2] The NRC-approved Fire Protection Program includes the fire protection and post-fire safe shutdown systems necessary to satisfy NRC guidelines and requirements; administrative and technical controls; the fire brigade and fire protection related technical staff; and other related plant features which have been described by the licensee in the FSAR, fire hazards analysis, responses to staff requests for additional information, comparisons of plant designs to applicable NRC fire protection guidelines and requirements, and descriptions of the methodology for assuring safe plant shutdown following a fire.

protection systems. The incorporation of the fire protection system, including impairments and their associated compensatory measures, into the TS was deemed necessary to assure that:

- Prompt compensatory measures would be taken,
- An appropriate alternate means of protection is provided, and
- Appropriate organizations are notified to counter the effects of systems/equipment being out-of-service.

In 1986, the Commission provided an approach for licensees to remove unnecessary fire protection Technical Specifications (TS). (GL 86-10). Under this guidance, most TS requirements for fire protection system operability (and associated compensatory measures) were relocated from the technical specifications to documents referred to in the FSAR, such as a technical requirements manual (TRM). Specific FPP elements that remained in the TSs include: (a) specifications associated with safe-shutdown equipment; and, (b) administrative controls which ensure that the NRC-approved FPP is appropriately established, implemented, and maintained in accordance with the current license basis (CLB). A review of several FPPs found that that relocating fire protection features from the TS to licensee controlled documents referenced in the FSAR did not result in a significant change to the operability requirements for fire protection features or their associated compensatory measures.

For most impairments, compensatory measures are specified in the NRC-approved FPP. The most common type of compensatory measure is a fire watch, which is a person trained to look for fire hazards, detect early signs of fire, and initiate alarms. For unique impairments not addressed by the FPP or where the compensatory measure specified in the FPP does not provide an appropriate level of protection for the specific circumstances, an alternate approach may be determined by the licensee in accordance with NRC guidance (RIS 2005-07). The evaluation should consider the impact of the proposed alternate compensatory measure to the FPP and its adequacy compared to the compensatory measure required by the FPP. In addition, the evaluation must demonstrate that the alternate compensatory measure would not adversely affect the ability to achieve and maintain safe shutdown in the event of a fire. For example, if an evaluation of the specific hazards in the area finds that sole reliance on a fire watch would not be sufficient to assure the ability to achieve and maintain safe shutdown conditions in the event of fire, a licensee would be expected to implement an alternate measure. Depending on the specific circumstances, alternate compensatory measures may be limited to enhancements of the compensatory measure specified in the FPP, such as additional administrative controls, operator briefings, temporary procedures, and interim shutdown strategies; or they may be more comprehensive such as procedures which specify actions to be taken by operators outside the control room in the event of fire, installing temporary fire barrier penetration sealing materials such as intumescent pillows and blocks and the use advanced fire protection technologies.

Compensatory measures are expected to *temporary*. They are not intended to be in place for extended periods of time or used as a technique to avoid the completion of activities required to achieve full compliance with regulatory requirements. Nevertheless, some NPPs have been found to rely on compensatory measures for extended periods of time (in some cases years). In 2005 the NRC informed licensees that it expects corrective action(s) will be completed, and reliance on the compensatory measure eliminated, at the first available opportunity, typically the first refueling outage (RIS 2005-20) . More recently, in a March 30, 2012 letter report (ML120900777), NRC defines a *"long term compensatory measure"* as one that has been in place for longer than 18 months. This means that the functionality[3] of impaired fire protection feature(s) is expected to be restored no later than 18 months from the date of discovery.

This report consolidates a number of NRC communications and technical documents related to the use of fire protection compensatory measures at commercial NPPs. In addition to providing an historical overview, the report details the current regulatory framework for compensatory measures and illustrates how methods different from those specified in a plant's approved fire protection program may be used to provide an effective alternative when unique circumstances are encountered. As such, the document is primarily intended to serve as a knowledge base for members of the NRC staff. However, NPP operators may also find it to serve as a useful reference.

---

[3] To be considered operable, a fire protection system or component that is not controlled by the plant's TS, does not have to be "fully qualified" in terms of its design and licensing bases as long as the licensee can demonstrate functionality. This definition does not apply to systems or components controlled by the plant's TS (Ref.51).

ACKNOWLEDGEMENTS

The author would like to thank the individuals who contributed their time and expertise in providing input and reviewing drafts of this report. In particular, Mr. Felix Gonzalez and Mr. Mark H. Salley of the Office of Nuclear Regulatory Research, and Mr. Phillip M. Qualls of the Office of Nuclear Reactor Regulation. Also, I greatly appreciate the comments and suggestions received from Mr. Dan Frumkin and Mr. Brian Metzger of the NRC Office of Nuclear Reactor Regulation; Mr. Patrick Finney, Mr. Keith Young, and Mr. John Rogge of NRC Region I; Mr. Gerald Wiseman of NRC Region II; and Mr. Robert Krsek and Mr. Dariusz Szwarc of NRC Region III. I would also like to express my gratitude for the assistance provided by my colleagues, Mr. Randy Weidner and Mr. Jim Higgins of BNL.

Finally, I would like to acknowledge the excellent assistance offered by Ms. Maryann Julian and Ms. Maria Anzaldi of BNL in preparing the document and Eric Goss of NRC's Graphics Department in preparing some of the Figures.

This report was issued for 45 day public comment on TBD, 2013 (FR XXXX). I would also like to acknowledge the following public comments:

ACRONYMS AND ABBREVIATIONS

ALARA	As Low As is Reasonably Achievable
AOP	Abnormal operating procedure
APCSB	Auxiliary Power Conversion Systems Branch
ASD	Alternate Shutdown or Aspirating Smoke Detection, depending on usage.
BNL	Brookhaven National Laboratory
BTP	Branch Technical Position
CFR	Code of Federal Regulations
CLB	Current Licensing Basis
CO_2	Carbon Dioxide
DID	Defense-In-Depth
EGM	Enforcement Guidance Memorandum
EPRI	Electric Power Research Institute
ERFBS	Electrical Raceway Fire Barrier Systems
FAQ	Frequently Asked Question
FHA	Fire Hazards Analysis
FPP	Fire Protection Program
FR	Federal Register
FSAR	Final Safety Analysis Report
GDC	General Design Criteria
GL	Generic Letter
HVAC	Heating, Ventilation and Air Conditioning
IEEE	Institute of Electrical and Electronics Engineers
IN	Information Notice
IP	Internet Protocol
LCIE	License Condition Impact Evaluation
LCO	Limiting Condition of Operation
LED	Light-emitting diode
LER	Licensee Event Report
NEI	Nuclear Energy Institute
NFPA	National Fire Protection Association
NPP	Nuclear Power Plant
NRC	Nuclear Regulatory Commission
NRR	Office of Nuclear Reactor Regulation (at NRC)
NUMARC	Nuclear Management and Resources Council
OE	Office of Enforcement

OMA	Operator Manual Action
PRA	Probabilistic Risk Assessment
QA	Quality Assurance
RES	Office of Nuclear Regulatory Research (at NRC)
RG	Regulatory Guide
RIS	Regulatory Issue Summary
SCBA	Self-contained breathing apparatus
SDP	Significance Determination Process
SER	Safety Evaluation Report
SRG	Special Review Group
SRM	Staff Requirements Memorandum
SRP	Standard Review Plan
SSA	Safe Shutdown Analysis
SSD	Safe Shutdown
SSC	Structures, Systems and Components
STS	Standard Technical Specifications
TS	Technical Specification
UFSAR	Updated Final Safety Analysis Report
VEWFD	Very Early Warning Fire Detection
VID	Video Image Detection

1. INTRODUCTION

1.1 Background

A fire which occurred at the Browns Ferry Nuclear Power Plant on March 22, 1975 led to significant changes in the fire protection regulatory framework for commercial nuclear power plants (NPP). As part of the Commissions response to the fire, the Executive Director for Operations of the NRC established a Special Review Group (SRG) to identify lessons learned and make recommendations for improving the level of fire safety at commercial NPPs. In its report, NUREG-0050, "Recommendations Related to the Browns Ferry Fire: Report by Special Review Group" [Ref. NUREG-0050], the SRG noted that when fire protection features are disabled temporary measures must be provided for fire protection in areas covered by the disabled equipment. This recommendation has been incorporated into current fire protection programs by requiring licensees to implement temporary actions, called *"interim compensatory measures,"* (or *"compensatory measures"*) to offset the reduction in fire safety caused by the impaired[4] fire protection feature until corrective actions are completed. Examples of common impairments include damaged or degraded fire barriers, inoperable smoke detectors, fire protection components taken out-of-service to facilitate maintenance, and broken hardware on fire-rated doors.

Based on recommendations contained in the SRG's report, Subsequent to the SRG report, in 1976, the NRC issued Branch Technical Position Auxiliary and Power Conversion Systems Branch 9.5-1 (BTP APCSB 9.5-1), "Guidelines for Fire Protection for Nuclear Power Plants," to describe acceptable methods for implementing General Design Criterion 3, "Fire Protection" of Appendix A to 10 CFR Part 50, "General Design Criteria for Nuclear Power Plants" (GDC 3)[5]. A key objective of the BTP and its Appendix A, was for every NPP to establish a fire protection program (FPP) that is based on the defense-in-depth safety concept of using multiple, independent, layers of protection to assure the health and safety of the public. The BTP further specified that administrative procedures be provided to ensure impaired fire protection systems are reviewed by appropriate levels of management and appropriate actions and procedures are implemented to assure adequate fire protection and reactor safety until the impairment is permanently corrected.

Integrating the defense-in-depth concept into fire protection resulted in the establishment of comprehensive FPPs at each plant that are designed to achieve an adequate balance among each of the following fire safety objectives:

---

[4] When a fire protection feature or system cannot perform its intended function, it is said to be "impaired" (Ref. RG1.189).

[5] BTP APCSB 9.5-1 did not apply to plants already in operation (i.e., docketed) at the time it was issued. Guidance for operating plants was provided later in Appendix A to BTP APCSB 9.5-1, " Guidelines for Fire Protection for Nuclear Power Plants Docketed Prior to July 1, 1976," which, to the extent practicable, relies on BTP APCSB 9.5-1.

1. Preventing fires from starting.
2. Detecting and extinguishing quickly fires that may start and limiting their damage.
3. Designing the plant to minimize the effect of fires on essential safety functions so the ability to achieve and maintain safe shutdown conditions will be preserved.

Impairments cause the overall level of fire protection defense-in-depth to be degraded. Therefore, their associated compensatory measures should

- be promptly implemented,
- provide an appropriate level of compensation, and
- be temporary; Reliance on the compensatory measure should be eliminated by the end of first available refueling outage

Examples of common compensatory measures include:

- fire watches

- enhanced controls over combustible materials and/or ignition sources,

- temporary repairs or modifications,

- temporary procedures

- operations crew briefings (e.g., Control Room Night Orders),

- staging temporary backup fire suppression equipment

As discussed in Regulatory Guide 1.189, "Fire Protection for Nuclear Power Plants" [RG 1.189] and Regulatory Information Summary (RIS) 2005-07: "Compensatory Measures To Satisfy The Fire Protection Program Requirements" [RIS 2005-07], implementing only those measures specified in the approved FPP may not be sufficient to mitigate fire risk in all circumstances. For example, if an evaluation of the specific hazards in the area finds that sole reliance on a fire watch would not be sufficient to assure the ability to achieve and maintain safe shutdown conditions in the event of fire, a licensee would be expected to implement an alternate measure or combination of measures. Depending on the specific circumstances, alternate compensatory measures may be limited to enhancements of the compensatory measure specified in the FPP, such as additional administrative controls, operator briefings, temporary procedures, and interim shutdown strategies; or they may be more comprehensive such as procedures which specify actions to be taken by operators outside the control room in the event of fire, installing temporary fire barrier penetration sealing materials such as intumescent pillows and blocks and / or the use advanced fire protection technologies. A licensee may implement alternate compensatory measures without prior approval of the NRC provided they are evaluated in accordance with the guidance provided by the staff in RIS 2005-07. However, changes to compensatory measures defined in the plant's Technical Specifications (TS) would require a license amendment.

1.2 Objective

During the development of the current fire protection regulatory framework, the NRC foresaw cases in which fire protection features would be inoperable, and allowed licensees to implement, in accordance with their approved fire protection program, interim measures to compensate for the impaired condition or component until permanent corrective actions were implemented. More recently, several generic communications have been issued by NRC staff to address various issues related to the application of compensatory measures for fire protection programs. For example, in 1997, NRC issued Information Notice (IN) 97-48 to alert addressees to potential problems associated with the implementation of interim compensatory measures. In Regulatory Issue Summary (RIS) 2005-07 (ML042360547) the staff describes how licensees may use alternate approaches, when the measures specified in the approved FPP do not provide an appropriate level of safety for the specific circumstances. Both IN 97-48 and RIS 2005-07 cite cases where compensatory measures specified in approved FPPs may not be appropriate for unique circumstances, such as degraded or inoperable components and fire protection features associated with the plant's post-fire safe-shutdown capability.

In 2009, the NRC Office of Nuclear Reactor Regulation (NRR) Fire Protection Branch Chief, recommended that longstanding concerns about compensatory measures be addressed and requested the Office of Nuclear Regulatory Research (RES) to prepare a document consolidating documentation regarding compensatory measures usage. Subsequently, RES contracted Brookhaven National Laboratory (BNL) to develop a reference guide for agency staff who evaluate the acceptability of interim compensatory measures when degraded or nonconforming conditions are identified in fire protection features at nuclear power plants.

1.3 Scope

BNL was requested to assist the NRC staff in the development of a NUREG series report which:

- provides an historical background of compensatory measures for fire protection;

- describes the fire protection regulatory framework concerning compensatory measures, including the relationship between compensatory measures and defense-in-depth.

- consolidates and clarifies existing NRC staff positions, interpretations, and guidance regarding compensatory measures

- describes the types of compensatory measures typically included in fire protection programs approved by the staff;
- identifies alternate approaches such as new fire protection technologies that may be used in lieu of, or as a supplement to, the originally approved compensatory measures; and,

- provides an illustrative example of how the use of new technologies may provide a more appropriate method of compensation than methods specified in the approved FPP when unique conditions exist.

This report is primarily intended to serve as a knowledge base for members of the NRC staff. It consolidates a number of NRC generic communications and technical documents related to the use of fire protection compensatory measures at commercial NPPs. However, it should be used as a compliment to, not a substitute for, requirements, inspection procedures, and standard review plans. Users are encouraged to review the references cited in the document to gain an in-depth understanding of compensatory measures.

2. DISCUSSION

2.1 Historical Overview

During the initial development of the U.S. nuclear reactor program in the early 1970s, fire protection requirements for commercial nuclear power plants were limited to the broad performance objectives of Appendix A to 10 CFR Part 50, *"General Design Criteria for Nuclear Power Plants."* Specifically, General Design Criteria 3, (GDC 3) specified that "structures, systems, and components important to safety be designed and located to minimize the probability, and adverse effects of fires and explosions, that noncombustible and heat- resistant material be used wherever practical, and that fire detection and suppression systems be provided to minimize the effect of fire on structures, systems, and components important to safety." Because of a lack of detailed implementation guidance, the level of fire protection adopted by a facility was generally found to be acceptable if it complied with local and industrial fire codes and received an acceptable rating from its fire insurance underwriter. Thus, the fire protection features at commercial NPPs were very similar to those of conventional, fossil-fueled, power generating stations.

A major fire at the Browns Ferry Nuclear Power Plant on March 22, 1975 underscored the need for more detailed criteria and guidance for satisfying GDC 3 (Ref. NUREG-0298). The revised regulatory framework which followed was largely based on recommendations made by a "Special Review Group" (SRG) appointed by the NRC Executive Director for Operations to identify lessons learned and make recommendations for improving fire safety. The main objective of the SRG was to (a) review the circumstances, origins and consequence of the fire and (b) propose improvements in NRC policies, procedures and technical requirements (Ref: NUREG-0050).

In its report (Ref. NUREG-0050), the SRG recognized the importance of providing *temporary enhancements* when fire protection features are disabled or taken out-of-service. Specifically, in its discussion of disabling fire protection systems for maintenance the SRG states: *"temporary measures" must be established for fire protection in areas covered by the disabled equipment."* This recommendation is reflected in the current fire protection program, which requires licensees to implement temporary actions, called *"compensatory measures"* (Ref. RG 1.189) to offset the reduction in fire safety caused by the impaired fire protection systems and features until corrective actions are completed and the component or system is capable of performing its specified design functions, as specified in the plant's current license basis (i.e., functionality[6] is restored).

Based on the SRG recommendations, the NRC staff developed comprehensive guidelines for satisfying GDC 3; they are documented in the following Branch Technical Positions (BTP):

---

[6] To be considered operable, a non-TS system or component does not have to be "fully qualified" in terms of its design and licensing bases as long as the licensee can demonstrate functionality. This definition does not apply to systems or components controlled by the plant's TS (Ref.51).

1. BTP Auxiliary Power Conversion Systems Branch (APCSB) 9.5-1, "Guidelines for Fire Protection for Nuclear Power Plants." BTP 9.5-1 was issued in May 1976 but was only applicable to plants which filed an application for construction after July 1, 1979,

2. Appendix A to BTP APCSB 9.5-1. Frequently identified as *"Appendix A to the BTP,"* these guidelines were issued in September 1976 for currently operating NPPs to offer acceptable alternatives in areas where strict compliance with BTP 9-5-1 would require significant modifications.

At the time these documents were issued, there was a wide variation in plant design, operating status and construction phase. Some plants had been operating for some time or in the final stages of construction, while others were in the early stages of design or construction. Many newer plants incorporated physical separation of safety systems into their designs. However, older plants, like Browns Ferry, had significantly less inherent separation. Thus, two separate guidance documents were deemed necessary to provide a balanced approach in the application of the SRG recommendations to specific facilities.

Following their promulgation, licensees were requested to evaluate their fire protection program against the appropriate guidelines. Toward the end of 1978, the NRC found that fifteen fire protection issues remained unresolved, with certain licensees refusing to adopt some of the recommendations in the new guidance. Even though only a few plants might contest a given issue, the Commission concluded that rulemaking was the appropriate vehicle for implementing their fire protection policy. Accordingly, in 1980, five years after the Browns Ferry fire, the Commission proposed a new fire protection rule to resolve the contested issues. In a letter dated November 24, 1980 (Ref. GL 80-100), the Commission informed all power-reactor licensees with plants licensed before January 1, 1979, of the new fire protection regulations in Section 10 CFR 50.48 ,"Fire Protection," and Appendix R to 10 CFR 50 "Fire Protection Program for Nuclear Power Plants Operating Prior to January 1, 1979". Plants licensed to operate after this date were required to satisfy similar criteria, as specified by the provisions of their operating license.

Because it was not deemed possible at the time (1980) to accurately predict the manner in which a fire would start and propagate, the new fire protection requirements employed a "deterministic" approach which considers a set of challenges to fire safety and then determines how they should be mitigated. The key assumptions of this approach are:

- An exposure fire[7] occurs in any single fire area; and,

- Structures, Systems, and Components (SSCs) not conforming to the specified protection criteria cannot perform their intended function.

One example of a deterministic criterion is the requirement that electrical raceway fire barrier systems (ERFBS) relied upon to protect post-fire safe shutdown related cables have a fire rating of either 1- or 3-hours depending on the availability of other fire protection features (Ref. RG 1.189). Due to deficiencies identified in the early 1990s with a widely used ERFBS manufactured by Thermal Science, Inc. (Thermo-lag 330-1), the staff requested that interim compensatory measures (i.e. fire watches), be implemented and remain in place until the licensee can declare the fire barriers operable (Ref. NRC Bulletin No. 92-01). In certain

---

[7] An *exposure fire* is a fire that involves either in situ or transient combustibles in a given fire area, and is external to any structures, systems, or components located in or adjacent to that same area.

instances, these interim compensatory measures remained in place for extended periods, in some cases years.

Recently, there have been significant advances in applying probabilistic risk assessment (PRA) techniques for fire protection at commercial NPPs. In 1998, the staff informed the Commission of its plans to work with the National Fire Protection Association (NFPA) and the nuclear power industry to develop a risk-informed, performance-based, fire protection consensus standard for NPPs (Ref. SECY-98-0058). In 2000, the staff received Commission's approval to proceed with rulemaking to permit reactor licensees to adopt NFPA 805[8] as a voluntary alternative to existing requirements (Ref. SECY 00-0009) On July 16, 2004, 10 CFR 50.48 "Fire Protection" was amended to add a new subsection, 10 CFR 50.48 (c) (Ref. 69 FR 33536) allowing licensees to adopt NFPA 805 on a voluntary basis. Hence, a licensee may choose risk-informed and performance-based fire protection requirements as an alternative to the deterministic requirements of 10 CFR 50.48(a) and Appendix R to 10 CFR 50. At present, approximately half of the operating NPPs have submitted letters of intent to transition to a risk-informed approach under 10 CFR 50.48(c).

Regardless of the selected approach, the goal of minimizing both the likelihood of occurrence and the consequences of fire in nuclear power plants remains the same. In both approaches, the requisite level of safety is assured through a "defense-in-depth" (DID) concept which incorporates multiple protective barriers to prevent the occurrence of fire, and features to mitigate its consequences should one occur. The aim is to provide a suitable balance between each layer of protection. Increased strength, redundancy, performance, or reliability of one layer can compensate in some measure for deficiencies in the others (Ref. NUREG 0050).

2.2   Purpose of Report

In 2008, the Commission directed the staff to formulate a closure plan for stabilizing the fire protection regulatory infrastructure (Ref. SRM-M080717). In response, the staff identified eight specific tasks for improving the regulatory stability of fire protection (Ref. SECY 08-0171). In a July 2009 Memo (Ref. Klein, 2009) the Chief of the Fire Protection Branch of NRR identifies Task 7 to consolidate regulatory documentation on compensatory measures and to identify available new technologies that potentially could be used in lieu of, or as a supplement to traditional measures specified in approved FPPs.

To address this issue, the NRC Office of Nuclear Regulatory Research (RES) contracted Brookhaven National Laboratory to assist them to prepare a NUREG series report that (a) provides a comprehensive "knowledge base" of regulatory and technical information about assessing compensatory measures for impaired fire protection elements; (b) describes the types of compensatory measures included in approved fire protection programs; and, (c) identifies new technologies that may offer a viable alternative to those delineated in approved fire protection programs.

RES envisioned that this NUREG report would be used by agency staff (inspectors and reviewers) who evaluates the acceptability of alternative compensatory measures after impaired conditions are identified in fire protection features at nuclear power plants. Therefore, the

---

[8] NFPA 805 is discussed for historical purposes only. Discussions on compensatory measures do not apply to NFPA 805 unless stated otherwise.

report's overall objective is to serve as a consolidated source of regulatory and technical information for the NRC staff responsible for assessing the appropriateness of fire-safety compensatory measures at commercial NPPs.

2.3   Scope of Report

This report provides guidance to assist the NRC staff in determining whether a compensatory measure is appropriate for a given impairment. It discusses the criteria and guidance upon which the approved fire protection program is based, and, to the extent practical, describes unique aspects of compensatory measures, such as the potential for advanced technologies to be implemented as an alternative to the "traditional" compensatory measures specified in the approved fire protection program. It should be noted that NFPA 805 is discussed for historical purposes only and discussions on compensatory measures do not apply to NFPA 805 unless stated otherwise.

3. REGULATORY FRAMEWORK

General Design Criterion 3 (GDC 3), *Fire Protection*, of Appendix A to 10 CFR Part 50 establishes the fundamental performance objectives for fire protection at NPPs. In response to the fire at the Browns Ferry NPP, the NRC codified criteria for satisfying GDC 3 in Title 10 of the *Code of Federal Regulations,* Part 50, Section 48, "Fire Protection" (10 CFR 50.48). Paragraph (a) of 10 CFR 50.48 (50.48(a)) requires, in part, that each operating nuclear power plant have a fire protection plan that describes the overall fire protection program, and specific features necessary to implement the program. In July 2004, the NRC amended 10 CFR Part 50.48 adding a new subsection, 10 CFR 50.48(c) to allow licensees to adopt, on a voluntary basis, a risk informed approach described in the 2001 Edition of National Fire Protection Association (NFPA) Standard 805[9], *"Performance-Based Standard for Fire Protection for Light Water Reactor Electric Generating Plants."* Regardless of the selected approach, all plants must develop and maintain a fire protection program (FPP) which provides assurance, through a defense-in-depth approach, that the Commission's fire protection objectives are satisfied. These objectives are:

1. Minimize the potential for fires and explosions to occur;
2. Rapidly detect, control, and extinguish fires that do occur; and,
3. Ensure that fire will not prevent the performance of necessary safe shutdown functions, and will not significantly increase the risk of radioactive releases to the environment (Ref. NUREG 0800).

The specific features (i.e., personnel, structures, systems, components, analyses, procedures and policies) which collectively form the FPP are described in licensee controlled documents, such as:

- Fire Hazards Analysis (FHA)
- Safe Shutdown Analysis (SSA)
- Plant operating procedures
- Surveillance, maintenance and test procedures
- Fire brigade response procedures and pre-fire plans

Since the Browns Ferry fire in 1975, the NRC has provided additional information on a number of fire protection issues. Guidance related to use of compensatory measures is principally contained in the following:

1. Appendix A to BTP APCSB 9.5-1, "Guidelines for Fire Protection for Nuclear Power Plants Docketed Prior to July 1, 1976

   Specifies that licensees should establish administrative controls for the impairment of fire detection and suppression systems and special actions and procedures, such as fire watches or temporary fire barriers put in place to ensure adequate fire protection and reactor safety.

---

[9] NFPA 805 is discussed for historical purposes only. Discussions on compensatory measures do not apply to NFPA 805 unless stated otherwise.

2. Generic Letter 81-12, "Fire Protection Rule," February 20, 1981

    In addition to defining safe-shutdown objectives, reactor performance goals, safe-shutdown systems and components, and Requested licensee's to develop TS's surveillance requirements and limiting conditions for operation (LCO) for safe shutdown equipment that was not addressed in the plant's existing TS.

3. Generic Letter 86-10, "Implementation of Fire Protection Requirements", April 24, 1986

    This Letter recommends that each licensee place its fire protection program and major commitments into its FSAR and, encourages licensees to:

    -   replace current license conditions regarding fire protection and adopt the *Standard Fire Protection License Condition* provided in the GL
    -   add administrative controls to TS consistent with those for other programs implemented by the license condition
    -   remove most fire protection requirements from the TS

    The GL also clarifies that temporary changes to specific fire protection features which may be necessary to accomplish maintenance or modifications are acceptable provided appropriate interim compensatory measures are implemented.

4. Generic Letter 88-12, "Removal of Fire Protection Requirements from Technical Specifications," August 2, 1988.

    Contains supplemental guidance for licensees preparing a license amendment request that conforms to GL 86-10. It identifies four major areas where the fire protection requirements should be relocated from the TS to the Fire Protection Program, and clarifies that the technical specifications associated with safe shutdown (SSD) equipment and the administrative controls related to fire protection audits were to be retained. As described in the GL, a conforming amendment would remove the fire protection requirements from the TS in four major areas:

    1. Fire detection systems,
    2. Fire suppression systems,
    3. Fire barriers, and
    4. Fire brigade staffing requirements.

5. Generic Letter 91-18. Revision 1, "Information to Licensees Regarding NRC Inspection Manual Section On Resolution of Degraded and Nonconforming Conditions," October 8, 1997.

    This Letter addresses licensee activities in resolving degraded and nonconforming conditions for plants that are at power and for those that will resume operations from any shutdown. Attachment 1, Inspection Manual, Part 9900 (IM 9900) Technical Guidance, "Resolution of Degraded and Nonconforming Conditions," identifies compensatory measures as an item to be considered in licensee assessments of reasonable assurance of safety for SSCs that are not expressly subject to TSs. In addition, IM 9900 provided guidance for evaluating compensatory measures as an interim step until final corrective actions were completed.

Note: As indicated in item 9 below, GL 91-18 was replaced by RIS 2005-20 and a 09/26/05 September 26, 2005 revision to NRC IM 9900, " Technical Guidance: Operability Determinations and Functionality Assessments for Resolution of Degraded and Nonconforming Conditions Adverse to Quality or Safety."

6. Information Notice 97-48, "Inadequate or Inappropriate Interim Fire Protection Compensatory Measures," July 9, 1997.

    This Notice was issued to alert licensees to potential problems associated with the implementation of interim compensatory measures for degraded or inoperable plant fire protection features or conditions associated with post-fire safe shutdown capability.

7. Response to Region IV Task Interface Agreement (TIA) (96TIA008) - Evaluation of Definition of Continuous Fire Watch, August 17, 1998.

    This Memorandum Documents the staff's evaluation of a fire protection inspection issue concerning the definition of a continuous fire watch. Specifically, the licensee's FPP defined a continuous fire watch as a fire watch that patrolled the fire area(s) of concern at least once every 15 minutes. Based on its evaluation, the staff disagreed, and concluded that a continuous fire watch is to remain in the affected fire area at all times.

    The TIA also notes that in addition to a fire watch, enhanced compensatory measures may be warranted to fully address potential safety issues presented by the nonconformance. Specific examples cited in the staff evaluation include enhancing controls over combustible materials and hot work, briefing operators and fire brigade members on the nonconformance condition, implementing temporary operating procedures, pre-staging manual firefighting equipment, and installing temporary fire protection features.

8. NRC Regulatory Issue Summary (RIS) 2005-07 – "Compensatory Measures to Satisfy the Fire Protection Program Requirements," April 19, 2005

    Informs addressees that under certain circumstances, compensatory measures other than those delineated in the approved fire protection program may be used to compensate for degraded or inoperable fire protection features. In addition, it provides specific guidance on how a licensee, with the GL 86-10 standard license condition for fire protection, may change the approved FPP to employ such alternate measures.

9. NRC Inspection Manual Part 9900, "Technical Guidance: Operability Determinations and Functionality Assessments for Resolution of Degraded and Nonconforming Conditions Adverse to Quality or Safety" September 26, 2005.

    This guidance was developed to assist NRC inspectors in their review of licensee determinations of operability and resolution of degraded or nonconforming conditions. This manual covers fire protection systems, structures, and components (SSCs) within the scope of the guidance for reviewing licensee actions for degraded and nonconforming conditions. . This version of IM 9900 supersedes guidance set out in Revision 1 of GL 91-18, "Information to Licensees Regarding Two NRC Inspection Manual Sections on Resolution of Degraded and Nonconforming Conditions and on Operability," dated October 8, 1997.

10. RIS 2006-10, "Regulatory Expectations with Appendix R Paragraph III.G.2 Operator Manual Actions," June 30, 2006:

RIS 2006-10 discusses the use of operator manual actions as a means of satisfying Section III.G of Appendix R to 10 CFR 50. Specifically, the RIS notes that some licensees relied on operator manual actions to compensate for degraded or missing fire barriers in lieu of providing one of the protective features specified in paragraph III.G.2 or requesting an exemption under 10 CFR Part 50.12, "Specific Exemptions."[10] With regard to compensatory measures for missing or degraded fire barriers, Section 2.5 "Compensatory Measures and Corrective Actions" states that for missing or degraded fire barriers, the following should occur:

- compensatory measures for missing or degraded fire barriers should be implemented, as required, in accordance with the licensees' approved fire protection program,
- missing or degraded fire barriers should be reported in accordance with the requirements of 10 CFR 50.72(b) (3) (ii) and 50.73(a) (2) (ii), and the guidance in NUREG-1022, "Event Reporting Guidelines 10 CFR 50.72 and 10 CFR 50.73, "
- missing or degraded fire barriers should be documented in accordance with their corrective action program including a detailed description of the affected structures, systems, or components (e.g., circuits)
- Corrective actions should be completed in accord with the guidance in RIS 2005-20, "Revision to Guidance Formerly Contained in NRC Generic Letter 91-18, Information to Licensees Regarding Two NRC Inspection Manual Sections on Resolution of Degraded and Nonconforming Conditions and on Operability."

11. Regulatory Guide 1.189, "Fire Protection For Nuclear Power Plants," Revision 2, October 2009

This Regulatory Guide contains comprehensive fire protection guidance and specific criteria for defining an acceptable FPP at NPPs. Section 1.5 details the regulatory position on compensatory measures for degraded and nonconforming conditions.

12. NFPA 805 "Performance-Based Standard for Fire Protection for Light Water Reactor Electric Generating Plants" 2001

Compensatory measures are termed "*compensatory actions*" and defined as:

> *"Actions taken if an impairment to a required system, feature, or component prevents that system, feature, or component from performing its intended function. These actions are a temporary alternative means of providing reasonable assurance that the necessary function will be compensated for during the impairment, or an act to mitigate the consequence of a fire. Compensatory*

---

[10] Plants licensed to operate on or after January 1, 1979 (post-1979 licensees), are not required to meet the requirements of paragraph III.G.2. Therefore, for this group of plants, a staff decision in an SER that approves the use of manual operator actions does not require exemption under 10 CFR 50.12. However, Post-1979 licensees may be requested to demonstrate, as part of the NRC Reactor Oversight Process, that the use of an operator manual action would not adversely affect the ability to achieve and maintain safe shutdown in the event of a fire consistent with their license.

measures include but are not limited to actions such as fire watches, administrative controls, temporary systems, and features of components."

NFPA 805 requires the development of procedures for compensatory actions to be implemented when fire protection systems and other systems credited by the fire protection program and this standard cannot perform their intended function and limit the duration of impairment. In addition, NFPA 805 specifies that compensatory actions should be appropriate with the level of risk created by the unavailable equipment, and that plant procedures should ensure that compensatory actions are not a substitute for promptly restoring the impaired system.

Major Documents/Milestones of FP and Compensatory Measures in the Nuclear Industry

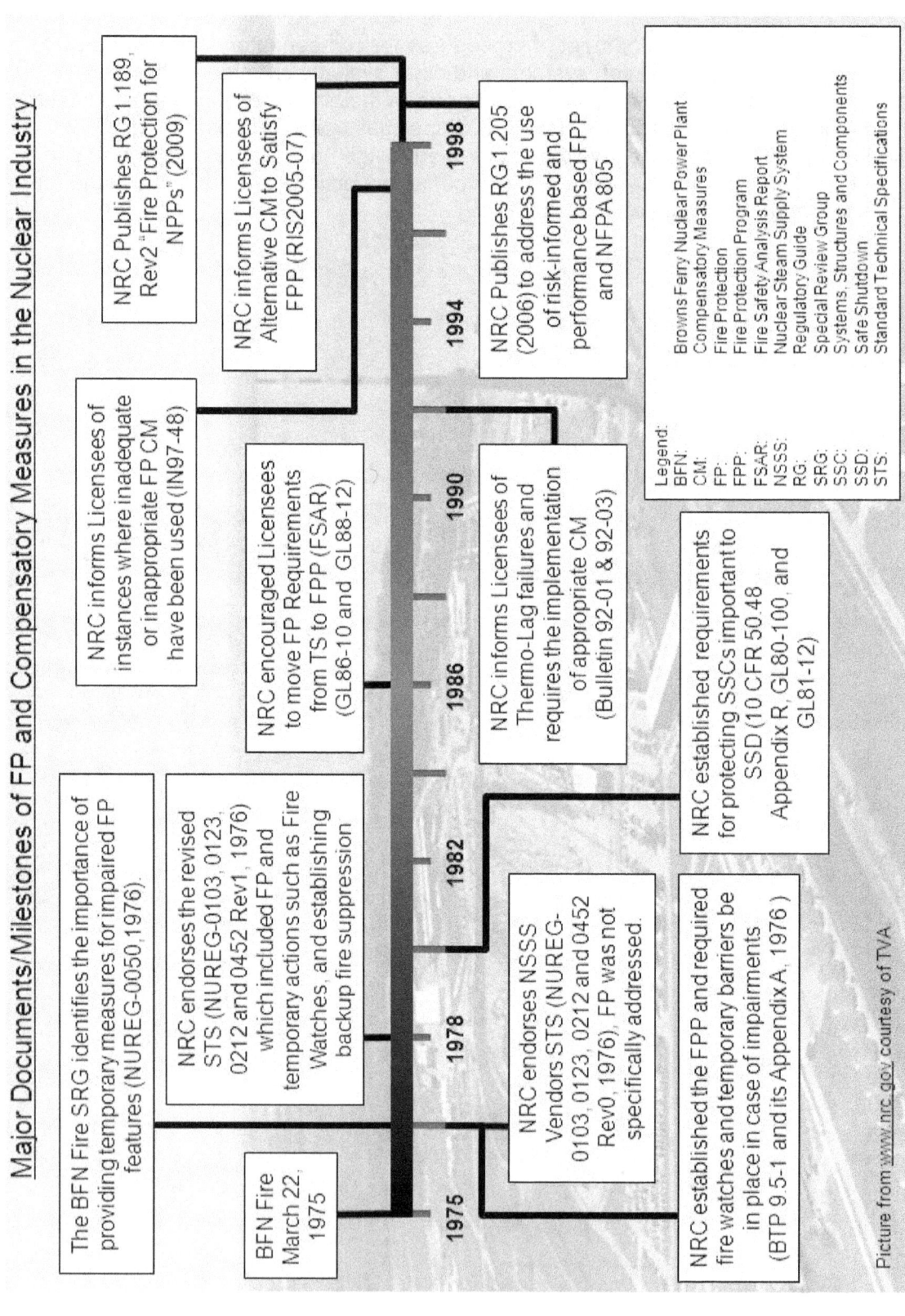

Figure 3-1. Major Documents and Milestones of FP and Compensatory Measures in the Nuclear Industry

3.1 Historical Overview of Fire Protection Technical Specifications

Technical specifications (TS) define the limits and conditions necessary to assure that the plant is operated in a manner consistent with the analyses and evaluations in the plant's Safety Analysis Report (Ref. 10 CFR 50.36). Specifically, the TS establish the following:

- Limiting Conditions For Operation (LCO) that constitute the lowest functional capability or performance levels of equipment required to safely operate the facility;

- Surveillances (e.g., tests or measurements) needed to confirm the operability of structures systems and components (SSCs) identified in the TS.

One of the first actions taken by the Commission in response to the Browns Ferry fire was to impose TSs for fire protection systems. During the 1976-1977 timeframe, each operating plant was provided a sample of recommended standard TS for fire protection and was requested to do the following:

a) Compare TS for existing fire protection systems against that sample; and,

b) Provide proposed fire protection TS for staff review.

The incorporation fire protection systems including impairments and their associated compensatory measures into the TS, was deemed necessary to assure that:

- Prompt compensatory measures would be taken,

- Appropriate temporary protection features would be provided, and

- Appropriate organizations are notified to counter the effects of systems/equipment being out-of-service.

FP systems, and features typically included in the scope of the TS include:

- Fire Protection Water Supply System

- Detection Systems

- Sprinkler / Spray Systems

- Fire Hose Stations

- Halon Systems

- CO_2 Systems

- Fire Barriers Penetration Seals

- Fire Brigade Staffing

- Rated fire barriers and their components (Fire doors, fire dampers, etc.)

This list subsequently was expanded by Generic Letter 81-12 to cover TS for safe shutdown equipment that was not already included in the plant's existing TSs.

On October 26, 1984, a fire protection policy steering committee (SC) recommended bringing together the existing fire protection guidance together in one generic letter and improving the consistency of information provided in the Standard Review Plan, TS, and operating licenses. Specifically, the SC recommended that the staff:

1. Develop and implement a standard fire protection license condition requiring compliance with the provisions of the Fire Protection Program, as described in the FSAR.

2. Revise the Standard Review Plan (SRP) to assure that Appendix R is fully covered

3. Revise the Standard Technical Specifications (STS) in a manner which assures functioning of fire protection features but which requires only those activities which are commensurate with other TS items in terms of their importance to safety (Ref. FP Policy Steering Committee Memo, 1984).

In GL 86-10 the Commission recommended that licensees apply for an amendment to their operating licenses to remove unnecessary fire protection Technical Specifications (TS). Additional guidance for the preparation of a license amendment request to implement Generic Letter 86-10 was provided on August 2, 1988 via Generic Letter 88-12, "Removal of Fire Protection Requirements from Technical Specifications." As discussed in GL 88-12, such an amendment (1) institutes the standard license condition for a Fire Protection Program. (2) removes requirements for fire protection systems from Technical Specifications (TS), (3) removes fire brigade staffing requirements from TS, and (4) adds administrative controls to TS that are consistent with those for other programs implemented by license condition.
By making fire protection program systems and features consistent with other plant features described in the FSAR, adoption of the "standard license condition" specified in GL 86-10, provides licensees greater flexibility in managing their fire protection programs. clarified To balance GL 88-12 further specified that the Administrative Controls section of the TS be augmented to support the Fire Protection Program by adding two specifications to the TS:

1. The Unit Review Group (Onsite Review Group) shall be given responsibility for reviewing the Fire Protection Program and implementing procedures, and for submitting recommended changes to the Company Nuclear Review and Audit Group (Offsite or Corporate Review Group).

2. The implementation of the Fire Protection Program implementation shall be added to the list of elements for which written procedures shall be established, implemented, and maintained.

In 1999 the Commission determined that the requirement to have the onsite review group (or a similar committee) review all changes to the fire protection program could place an unnecessary burden on licensees and, issued Regulatory Issue Summary (RIS) 99-002 "Relaxation of the Technical Specification Requirements for PORC Review of Fire Protection Program Changes" to allow a licensee to request a license amendment to remove the TS requirement provided it can be demonstrated that adequate controls are in place to ensure the maintenance of the effectiveness of their fire protection program.

Under the guidelines of GL 86-10 and 88-12, the requirements for fire protection operability and their associated compensatory measures were relocated from the TS to documents referred to in the FSAR (e.g., administrative procedures and / or technical requirements manual). Consequently, the operability requirements of most fire protection features that formally were included the plant's TS currently are governed by licensee controlled procedures referenced in the NRC-approved fire protection program. Specific FPP elements that remain in the scope of current TSs include: (a) specifications associated with safe-shutdown equipment; and, (b) administrative controls which ensure that the NRC-approved FPP is appropriately established, implemented, and maintained in accordance with the current license basis (CLB).

Based on a review of a sample of approved FPPs, the relocation of fire protection features from the TS to documents/procedures referenced in the CLB, does not appear to have had a significant impact on either the operability requirements for fire protection features or their associated compensatory measures. Appendix A summarizes the results of this review.

3.2   Fire Protection Program

The NRC-approved Fire Protection Program includes the fire protection and post-fire safe shutdown systems necessary to satisfy NRC guidelines and requirements; administrative and technical controls; the fire brigade and fire protection related technical staff; and other related plant features which have been described by the licensee in the FSAR, fire hazards analysis, responses to staff requests for additional information, comparisons of plant designs to applicable NRC fire protection guidelines and requirements, and descriptions of the methodology for assuring safe plant shutdown following a fire. (GL 88-12)

3.2.1   Design Features

Implementation of the defense-in-depth safety concept required each NPP to employ a wide array of plant design features that are directed at: (a) detecting, containing, and rapidly suppressing fires that may occur (in spite of fire prevention efforts) and (b) preventing fire from damaging SSCs important to safety. Common fire protection design features include:

- Structural fire barriers (e.g., fire-rated doors, fire-rated HVAC dampers, fire walls and cable penetrations)

- Electrical raceway fire barrier systems (e.g., ERFBS)

- Fire detection and alarm systems

- Fire suppression systems and equipment, including:

  - Fire water supply systems
  - Deluge water spray systems
  - Wet pipe sprinkler systems
  - Pre-action sprinkler systems
  - Dry pipe sprinkler systems
  - Carbon dioxide systems
  - Halon 1301 systems
  - Dry chemical suppression systems
  - Foam suppression systems
  - Manual firefighting equipment

- Post-fire safe shutdown capability, including:

  - SSCs required to perform essential shutdown functions (e.g., reactivity control, reactor coolant makeup, reactor pressure control, and decay heat removal)
  - Alternate Shutdown Panels
  - Emergency lighting and communications

The design of fire protection systems and equipment should satisfy criteria specified in the plant's fire protection license basis documents, such as:

- Final Safety Analysis Report (FSAR)

- Various Institute Of Electrical And Electronics Engineers(IEEE) Standards, such as IEEE Standard 634 -1978 "Standard Cable Penetration Fire Stop Qualification Test"

- Various National Fire Protection Association Codes, such as NFPA 13 - "Standard for the Installation of Sprinkler Systems"

- American Society for Testing and Materials (ASTM) E119-1976 "Fire Test of Building Construction and Materials"

3.2.2   Administrative Controls

In addition to design features needed to provide the requisite high level of fire safety, the defense-in-depth approach requires appropriate administrative controls be established to ensure the integrity of the FPP and equipment is maintained over the life of the plant. In addition, the Administrative Controls section of the plant's technical specifications typically requires the FPP to be established, implemented, and maintained in accordance with written procedures and administrative policies. With regard to compensatory measures, procedures should be established to identify and control fire protection system impairments and ensure appropriate remedial actions are taken. Additional examples of administratively controlled elements of the FPP include:

- The fire protection organization, including staffing requirements and responsibilities
- Transient combustible control
- Ignition source control
- Review and control of modifications to ensure that fixed fire loadings are not increased beyond those accounted for in the fire hazards analysis, or, if increased, that suitable protection is provided and the fire hazards analysis is revised accordingly.
- Review and control of temporary modifications and maintenance.
- Installation and use of temporary electrical power services
- Periodic testing and inspection of fire protection features

Additional specific guidance on administrative controls is provided in Section 2 of RG 1.189.

3.2.3   Quality Assurance

The quality assurance (QA) program for fire protection ensures that inspection, surveillance, and tests that govern the FPP are prescribed by documented instructions, procedures, and drawings. The QA program also ensures that testing is performed to demonstrate conformance with design and system readiness requirements. Thus, the QA program provides assurance that the fire protection systems and features are designed, fabricated, erected, tested, maintained, and operated so that they function as intended when needed.
Since fire detection and suppression systems are not safety-related they are generally not included within the scope of 10 CFR Part 50, Appendix B, "Quality Assurance Criteria for Nuclear Power Plants and Fuel Reprocessing Plants," unless the licensee has committed to include these systems under the plant's Appendix B program. Licensees that have not committed to include fire protection systems within the scope of their Appendix B program must provide a description of the fire protection QA program and the measures for implementing it for the NRC's review. For plants licensed after January 1, 1979, this review is conducted in accordance with the guidance provided in Regulatory Guide 1.189, "Fire Protection For Nuclear Power Plants," Revision 2, (RG 1.189). The QA program for fire protection at plants licensed before January 1, 1979, is reviewed against similar guidance contained in Appendix A to

APCSB 9.5-1 and Generic Letter 77-02, "Nuclear Plant Fire Protection Functional Responsibilities, Administrative Controls and Quality Assurance," (GL 77-02).

Examples of fire protection QA program features related to compensatory measures include:

- Procedures which control plant activities affecting fire protection equipment important to safety.

- Reporting, tracking, and ensuring restoration of fire protection equipment or features which have become inoperable

- Ensuring that inspections and tests of fire protection equipment and features are performed in accordance with documented instructions, procedures, and drawings.

- Periodic inspections of materials subject to degradation, such as fire barriers, stops, seals, and fire retardant-coatings to ensure these items have not deteriorated or been damaged.

4. IMPAIRMENTS AND COMPENSATORY MEASURES

4.1 Fire Protection Impairments[11]

As discussed above, the fire protection program is composed of a number of individual elements (e.g., plant design features, administrative controls, trained personnel, and engineering analyses) that reduce the likelihood of fires and explosions, rapidly detect and suppress those fires that may occur, and ensure the capability to achieve and maintain safe shutdown conditions in the event of a significant fire. When one element of the FPP can not perform its intended function, the overall level of defense-in-depth is reduced and, therefore, the potential for loss from fire is increased.. Although FPP features are well designed and highly reliable, it is expected that the performance of certain features will degrade over the life of the plant. If a fire protection feature is not capable of performing its specified function, it is considered to be impaired. Examples of common impairments are shown in Table 4-1 below.

Table 4-1. Examples of Fire Protection Impairments

Fire Suppression Water Supply Systems
Inoperable fire pump
Water supply volume is less than system requirements.
Inadequate Flow Path (e.g., inoperable valve or pipe necessary to supply fixed suppression systems or hose station hydrants, automatic valves in the flow path do not actuate to their correct positions on a simulated actuation signal).
Sprinkler, Spray and Deluge Systems
Inoperable or degraded / impaired sprinkler heads.
Inoperable fire detection actuation system (If fire detection that automatically initiates a fire suppression system is inoperable, the fire suppression system is also considered to be inoperable).
SSCs that that interfere with the ability of fire suppression systems to perform their intended function (e.g., scaffolds, work platforms, temporary equipment, etc.).
Gaseous Fire Suppression Systems (CO2 and Halon)
Opening in a suppression system containment boundary (e.g., fire barrier wall).
Underweight or under-pressure halon cylinder / banks.
Area ventilation dampers fail to close upon receipt of a simulated actuation signal.
System isolation to support maintenance activities.

---

[11] As used in this document, phrases such as *"fire protection impairment," "fire protection element"* or *"fire protection feature"* are intended to include any component or feature identified NRC approved fire protection program documents (e.g., FSAR, FHA, SSA, TRM etc.). Thus, the phrases include impairments to both classic fire protection features (e.g., sprinkler systems, hose stations, etc.) and features provided to ensure post-fire safe shutdown capability (e.g., electrical raceway fire barrier wraps, alternate shutdown panel, etc.).

Table 4-1. Examples of Fire Protection Impairments

Fire Hose Stations
One or more of the fire hose stations inoperable.
SSCs that (e.g., scaffolds, work platforms, temporary equipment, etc.) that impede access to a fire hose station.
Degraded or missing gaskets in fire hose couplings.
Failed hose hydrostatic test.
Yard Fire Hydrants and Hydrant Hose Houses
One or more of the yard fire hydrants or associated hydrant hose houses are inoperable and are the primary means of fire suppression.
Fire Rated Assemblies (Electrical Raceway Fire Barrier Systems, Structural Fire Barriers and Penetration Seals)
One or more of the required barriers non-functional.
A breach of a fire rated assembly/barrier that forms the compartmentalization (e.g., fire areas, fire zones) boundary.
A breach of a fire rated assembly/barrier that provides fire proofing for structural steel that supports fire barriers.
As-installed fire barrier penetration seal differs from the as-designed condition (e.g., unsealed penetrations, gaps greater than acceptance criteria).
Fire Detection
The number of operable fire detection instruments is less than the minimum required.
False / spurious alarms.
Safe (III.G.2) and Alternative (III.L) Shutdown
A required component is inoperable or has been removed from service
A required support system (e.g., electrical power sources, remote control circuits, room cooling etc.) is impaired or inoperable (.
One or more designated Emergency Lighting units impaired (e.g., lamps fail to illuminate on loss of ac power, ELU not in correct position to illuminate intended target(s)).
Fire Doors
Door hardware loose or damaged (e.g., loose hinges, missing /damaged door handle, door latch stuck in retracted position).
Door or frame structurally damaged.
Fire door propped / blocked open to support maintenance or prevented from fully closing by equipment (e.g. running a temporary cable under the door to facilitate maintenance).
Fire Dampers
Damper or frame structurally damaged.
Operation of damper impaired by rust, corrosion, paint, dirt, etc.
Damaged or missing parts on damper release mechanism.

4.2   Compensatory Measures

When an impairment is identified, either an *operability determination* or *functionality assessment* is performed.[12] In general, a determination of operability is required when an SSC described in the plant's TSs cannot perform the safety function specified in the plant's design basis. For non-TS SSCs that are risk significant and/or perform functions specified in the plant's current licensing basis (CLB), a functionality assessment is appropriate. Functionality assessments are performed to ensure that the availability and reliability of SSCs are maintained. Implicit in this definition is the assumption of the full capability of all necessary attendant instrumentation, controls, normal and emergency electrical power sources, cooling or seal water, lubrication or other auxiliary equipment required for the performance of the system, subsystem, train, component, or device to complete their related support function(s).   Under the guidance of GL 86-10 and GL 88-12, technical specifications of fire protection program elements other than those related to the capability for safe shutdown following a fire, were relocated from the TS to documents referenced in the approved FPP (e.g., UFSAR, FHA, and Technical Requirements Manual).   Therefore, the majority of fire protection impairments are evaluated for functionality. Functionality is normally assessed and documented through the corrective action process.

When a fire protection feature is not capable of performing the function(s) specified in the FPP, appropriate compensatory measures should be promptly implemented to restore, in some measure, the reduction in defense-in-depth created by the degraded, inoperable, or nonconforming condition. Employing compensatory measures, on a short-term basis, is an integral part of NRC-approved fire protection programs. However, compensatory measures are not expected to be in place for an extended period of time. As discussed in RIS 2005-20 (Ref. RIS 2005-20), the NRC staff expects that the corrective action(s) will be completed, and reliance on the compensatory measure eliminated, at the first available opportunity, typically the first refueling outage. More recently, in a March 30, 2012 letter report (ML120900777), NRC defines a *"long term compensatory measure"* as one that has been in place for longer than 18 months. This means that the functionality[13] of impaired fire protection feature(s) is expected to be restored no later than 18 months from the date of discovery. Thus, a *"long-term compensatory measure"* is considered as one that is in place for more than 18 months. As discussed in RIS 2005-20, the NRC staff may approve the use of compensatory measures for longer periods if appropriately justified. In making its determination, the NRC staff considers safety significance, the effects on operability, the significance of the degradation, and what is necessary to implement the corrective action. They also may consider the time needed for designing, reviewing, approving, or procuring the repair or modification; the availability of specialized equipment for the repair or modification; and, whether the plant must be in hot or cold shutdown to implement the actions. If the licensee does not resolve the degraded or

---

[12] Detailed guidance on operability, functionality, and resolution of degraded and nonconforming conditions are provided in NRC Regulatory Issue Summary 2005-20, Revision 1, "Revision to NRC Inspection Manual (IMC) Part 9900 Technical Guidance, Operability Determinations & Functionality Assessments for Resolution of Degraded or Nonconforming Conditions Adverse to Quality or Safety" (Ref. RIS 2005-20).

[13] To be considered operable, a fire protection system or component that is not controlled by the plant's TS, does not have to be "fully qualified" in terms of its design and licensing bases as long as the licensee can demonstrate functionality. This definition does not apply to systems or components controlled by the plant's TS (Ref.51).

nonconforming condition at the first available opportunity, or does not appropriately justify a longer completion schedule, the staff would conclude that corrective action was not timely and would consider taking enforcement action.

4.2.1 Types of Compensatory Measures

For typical fire protection system impairments (e.g., inoperable fire detection or suppression systems) the appropriate compensatory measure(s) to be implemented are specified the plant procedures or other documents referenced in the approved FPP. Examples of common compensatory measures are shown below in Table 4-2. Occasionally, unique circumstances may be encountered that were not considered in the approved FPP. When such instances arise, licensee's are expected to determine appropriate compensatory measures on a case-by-case basis. The selected measures should be appropriate for the equipment or issue for which they were meant to compensate, and be consistent with the plant's license basis.

Table 4-2. Common Types of Compensatory Measures

Compensatory Measure	Description
Continuous Fire Watch	A continuous fire watch is an individual the serves as an uninterrupted fire watch in a single fire area. If all parts of the single fire area are not in the line of sight from a fixed watch station (e.g., line-of-sight vision is obstructed by equipment), the fire watch is to maintain watch over the entire area by patrolling the assigned fire area.
Hourly Fire Watch	An hourly (roving) fire watch is an individual assigned to observe posted area(s) 24 times in 24 hours, at 60 minute intervals. A roving once-per-shift fire watch patrol assigned to tour specific fire areas once every eight hours has been approved by the staff for certain specific circumstances such as impairments located inside containment.
Backup Suppression	A backup means of suppression that is provided to compensate for an impaired fire suppression system. Examples include backup pumping capability, supplemental water source(s), and additional lengths of fire hose.
Standard Video Monitoring	Use of standard closed-circuit television (CCTV) systems such as those used for plant security to enhance the level of fire protection provided by a roving fire watch when unique conditions exist.
Temporary Repairs	A short-term, temporary plant change to restore the functional capability of an impaired FPP feature. Examples include installing temporary fire barriers, penetration seal materials, and i emergency lighting units.
Enhanced Combustible Controls	Placing additional limits on the amount and/or type of combustible material materials in the area(s) of concern.
Enhanced Ignition Source Controls	Placing additional limits on the amount and/or type of ignition sources in the area(s) of concern.

Fire Watch

A properly trained fire watch strengthens the first element of fire protection defense-in-depth (fire prevention) by looking for fire hazards and other conditions that could lead to a fire. In addition, it offers detection capability and, depending upon plant-specific protocols[14], may initiate manual fire suppression activities. These actions all strengthen the second principle of fire protection defense-in-depth (rapid fire detection and suppression).

Fire watches typically are assigned when one of the following conditions exists:

1. Fire hazards involving burning, welding, or similar operations that can initiate fires or explosions (hot work).

2. When FPP systems or features are no longer capable of performing their intended function, as specified in the plant's fire protection license basis.

Specific fire watch training should provide appropriate instruction on fire watch duties, responsibilities, and required actions for the different types of fire watches, such as continuous, hot-work, and roving fire watch patrols. If applicable, fire watch qualifications should include hands-on training on a practice fire with the extinguishing equipment to be used while on fire watch. If fire watches are to be used as compensatory actions, the training should include recordkeeping requirements (Ref. RG1.189).

Personnel required to perform fire watch duties for degraded FPP features typically receive training in the following topics:

*   Purpose and responsibilities of a fire watch

*   Types and classification of fires

*   Selection and use of portable fire extinguishers

*   Actions to be taken upon discovering a possible fire

*   Types of combustibles and potential ignition sources

*   Conduct of plant rounds

*   Maintaining complete, accurate logs and records

Fire watches provided for hot-work[15] duties have the authority to stop these operations if unsafe conditions develop; these personnel typically are trained in the following:

*   The inherent hazards of the work site, and of the hot work.

*   Ensuring that safe conditions are maintained during hot work operations.

---

[14] Whether or not a fire watch engages in manual fire fighting activities is determined by the individual licensees'.

[15] Additional guidance for the use of hot work fire watches is provided in NFPA 51B, "Standard for Fire Prevention During Welding, Cutting, and Other Hot Work."

- Using fire-extinguishing equipment
- Procedures for sounding an alarm if there is a fire.

The hot-work fire watch is not a replacement for proper planning to prevent conditions that allow a fire to develop, regardless of the fire-fighting equipment available and the capabilities of the individuals involved (Ref. NFPA 51B, 2009).

In general, fire watches are responsible for the following:

- Knowing the location of the nearest in-plant communications device (e.g. GAI-TRONICS Box) in relation to the fire watch post.
- Being in the assigned watch area as required.
- Maintaining alertness during the assigned fire watch.
- Ensuring the completeness and accuracy of fire watch records
- Knowing how to properly report a fire to the Main Control Room (MCR).
- Having a clear understanding of the Fire Watch duties and post assignment.

A fire watch is currently the most common type of compensatory measure. The physical presence of a fire watch enhances fire prevention through prompt recognition and disposition of fire hazards. In the event a fire were to occur, the fire watch helps to assure that it would be detected and extinguished during its early stages.

The benefits of a fire watch have been described by the NRC's Office of Nuclear Reactor Regulation (NRR) as follows:

> *A fire watch provides more than simply a detection function. Personnel assigned to fire watches are trained by the licensee to inspect for the control of ignition sources, fire hazards, and combustible materials; to look for signs of incipient fires; to provide prompt notification of fire hazards and fires; and to take appropriate action to begin fire suppression activities. Fire watch personnel are capable of determining the size, the actual location, the source, and the type of fire—valuable information that cannot be provided by an automatic fire detection system. (Ref. 61FR16011)*

A roving fire watch patrol is typically assigned to observe posted area(s) 24 times in 24 hours, at 60 minute intervals. The patrol does not have to begin on the hour, but it must be completed within 60 minutes. However, a once-per-shift roving patrol that tours assigned areas once every eight hours has been approved by the staff for certain specific circumstances such as impairments located inside containment. In most cases, an hourly fire watch can effectively patrol multiple fire areas within the time available to complete each fire watch tour (i.e., 60 minutes). In some cases, however, the licensee may need to strategically to post several fire watches to assure the tours will be successfully completed in the allotted time-frame. Whether or not a single patrol can cover multiple fire areas varies with plant-specific conditions such as the size and complexity of the fire areas to be patrolled and their accessibility.

As its name implies, a continuous fire watch provides an uninterrupted watch of a single fire area. Task Interface Agreement (TIA) (Ref. NRR 96TIA0008, 1998)A procedure that does not require the fire watch to remain within the specified "fire area" at all times is not acceptable.

A continuous fire watch may, however, be assigned to monitor several fire zones or rooms that are located within a single fire area, provided that the zones or rooms are readily accessible and easily viewed by a single fire watch at a frequency of about every fifteen minutes, with a margin of five minutes. This approach is acceptable because the "fire area" would be continuously monitored. As described by the staff in a 1998 Task Interface Agreement (TIA) (Ref. NRR 96TIA0008), a continuous fire watch is defined as follows:

> *Continuous Fire Watch – A fire watch who performs a continuous watch over an assigned area(s), room(s), or object(s). All parts of the area may not be in the line of sight from a fixed watch station due to location of equipment, therefore the fire watch is to maintain watch over the entire area by patrolling the assigned area. The continuous fire watch can be assigned to watch more than one room as long as the rooms are in the same general area; the fire watch remains in the same general area; welding, grinding or burning is not in progress within the area; and the assigned patrol is made every 15 minutes.*

Personnel assigned to a continuous fire watch should have no additional assigned duties that would require their leaving the post. In addition, a continuous fire watch should not leave the assigned post unless it is terminated by operations (typically the shift supervisor); this includes drills and emergencies, unless plant conditions are too hazardous to remain.

Although a fire watch is the most common, in certain cases it may not provide a sufficiently *effective* type of compensatory measure. For example, NRC inspectors conducting a fire protection inspection in the mid-1990s, (Ref. IR50-458/97-201) identified a licensee that had established an hourly roving fire watch as the only compensatory measure for its removal major sections of electrical raceway fire barrier systems (ERFBS) used to separate redundant trains of shutdown cables. Because the licensee considered the complete removal of the ERFBS as being equivalent to a degraded barrier, an hourly fire watch was deemed an acceptable compensatory measure. However, as noted by the inspection team, the complete removal of ERFBS is not the same as a degraded ERFBS that might have cracks or does not meet its specified fire-resistance rating. Had degraded ERFBS remained in place, the fire damage would have likely been delayed long enough to permit the fire brigade to respond and extinguish the fire. The sole use of a fire watch for a safe shutdown function which is not adequately protected against fire damage is an inappropriate application of a compensatory measure. As discussed by the staff in Information Notice (IN) 97-48, reliance on a fire watch without alerting operators of the deficient post-fire safe shutdown equipment, or providing interim shutdown strategies, could force them to undertake complex trouble-shooting and repair activities to compensate for the post-fire, safe-shutdown-design deficiencies. A Fire watch enhances the level of protection afforded by degraded fire barriers; it was never intended to serve as a replacement for a fire barrier.

Temporary Repairs and Modifications

In certain cases, a minor modification can restore the functional capability of an impaired FPP feature. For example, the functionality of a breached structural fire barrier may be temporarily restored by plugging the gap with intumescent pillows or blocks such as those shown in Appendix B. Other examples include:

- Temporary emergency lighting, such as portable lighting units having an 8-hour capacity.

- staging backup suppression equipment, such as fire extinguishers, backup fire pumps, or additional lengths of fire hose

- temporary procedure changes, such as revising abnormal operating procedures (AOPs) to temporarily direct operators to use of feasible and reliable manual actions in the event of fire in areas where the impaired electrical raceway fire barriers are located

- use of portable (temporary) detection systems or cameras

- temporary modifications to hardware, such as removing motive and control power to preclude a fire-induced spurious operation of equipment

Temporary modifications such as these have been shown to provide an effective compensatory measure. As with all such measures, their selection, preparation, installation, and maintenance should be governed by established procedures which ensure that they are appropriate for the particular impairment and properly installed and maintained. However, as stated previously, a compensatory measure is not a permanent replacement for a regulatory requirement.

4.2.2   Selecting Appropriate Compensatory Measures

The selection of a specific type of compensatory measure for a given impairment should be based on its ability to offset the degradation in defense-in-depth created by the inoperable, degraded, or nonconforming condition (Ref. NRR 96TIA0008, 1998). Depending on the plant-specific circumstances, certain impairments may have a greater impact on defense-in-depth than others. For example, because it impacts both the detection and mitigation elements of defense-in-depth, fire detectors used to actuate fire suppression systems typically represent a more critically important component of a plant's fire protection program than detectors installed solely for early fire warning and notification. Therefore, the appropriateness of a proposed compensatory measure should be evaluated by a qualified fire protection engineer. . In certain cases, this assessment may determine that additional enhancements are needed to supplement the measure specified in the FPP (Ref. NRR, 96TIA0008, 1998). Some examples include:

- creating temporary "combustible free zones"

- temporarily prohibiting hot-work in the affected fire area

- adjusting work planning to minimize the introduction of combustibles or ignition sources that could increase the likelihood and severity of a fire

- installing closed circuit television (CCTV) capability in high radiation areas

- installing temporary detection and alarm systems

- informing fire brigade members of nonconforming suppression systems so appropriate pre-fire strategies can be developed

- pre-staging manual firefighting equipment

- briefing operators on potential impacts to safe shutdown capability

- temporary procedure changes

The evaluation also should consider the effects of the compensatory measure on other aspects of the facility. As an example, a licensee may plan to close a valve as a compensatory measure to temporarily isolate a leak. While this action may resolve the degraded condition from a fire protection perspective, it may also affect flow distribution to other components or systems, complicate operator responses to normal or off-normal conditions, or have other effects that should be reviewed (Ref. RG 1.189).

4.2.2.1 Examples of Inappropriate Compensatory Measures

Compensatory measures are routinely evaluated by the NRC as part of its inspection program. The majority of these reviews have found compensatory measures to be appropriate for the given impairment. . As discussed in the following examples, however, certain methods have not been found to provide an appropriate level of compensation. These examples are provided to illustrate the influence ineffective compensatory measures may have on defense-in-depth.

<u>Sole Reliance on Fire Watches for Impaired Post-fire Safe Shutdown Functions</u>

In 1996 an NRC inspection found that a licensee had relied solely on roving, hourly, fire watches as a compensatory measure for a number of post-fire safe-shutdown design deficiencies while long-term corrective actions were being evaluated (Ref.: Inspection Report 50-255/ 96-004, Accession No. 9605290105). Although the design issues could impact the operators ability to achieve and maintain safe shutdown in the event of fire, the roving fire watches were not supplemented by any additional measures, such as alerting the operators to deficient conditions or providing them with interim shutdown strategies for successfully dealing with fire damage to safe-shutdown components. As a result, placing and maintaining the plant in a safe shutdown condition following a fire in the affected fire areas would have required operators to perform a significant number of trouble-shooting, and repair activities to compensate for the design deficiencies. As noted by the Commission in its Notice of Violation (Ref. Significant Enforcement Action EA-96-131), reliance on roving one-hour fire patrols as interim compensatory measures for several safety significant design deficiencies indicated a lack of sensitivity to implementing immediate corrective actions commensurate with the safety significance of the deficiencies. .

<u>Improper Definition of a Continuous Fire Watch</u>

A continuous fire watch is an uninterrupted fire watch that is posted in a single fire area (Ref.: US NRC, 96TIA0008, 1998). Where there is no installed automatic fire detection system, the continuous presence of a fire watch is typically required to adequately compensate for the degraded, inoperable, or nonconforming condition. Depending on the size of the fire area and the specific hazards involved, it may be necessary to strategically post several continuous fire watches in single fire area to effectively maintain confidence that potential fire conditions will be promptly detected and reported.

An inspection conducted by NRC Region IV in 1995, found a licensee had revised its fire watch procedure to allow a single continuous fire watch to patrol multiple fire areas. Since the previous version of the procedure specified that a continuous fire watch was restricted to a single fire area, NRC Region IV asked the Office of Nuclear Reactor Regulation (NRR) staff to review the adequacy of the licensee's revised criteria. In its response (Ref. NRR 96TIA0008, 1998), the staff concluded the following:

- The definition of a continuous fire watch used by the licensee is not consistent with the intent of the continuous fire watch (to remain in the affected fire area at all times).

Draft for Comment

- The licensee did not provide sufficient technical justification for redefining the criteria for a continuous fire watch or for extending the frequency of the fire watch patrols and did not establish that the newly defined fire watch is equivalent to the previously defined fire watch.

- To the extent that other licensees have adopted similar definitions for continuous fire watches without adequate technical justification, our conclusions would be the same.

Inappropriate Elimination of a Compensatory Measure (Fire Watch)

In October of 2009, NRC Region IV inspectors observed maintenance personnel performing weld preparation work on essential service water piping using an abrasive flapper wheel grinder. Sparks were observed to extend four to five feet from the work area. However, there was no fire watch posted. When the inspectors questioned the maintenance personnel about this omission, they stated that since they were using a flapper wheel a fire watch was not required. From a subsequent review of applicable procedures, the inspectors determined that a prior (December 2003) procedure revision inappropriately eliminated the need to post a fire watch when using wire brushes, flapper wheels, polishing devices, or buffing pads mounted on power grinder motor drives or air tools. The inspectors determined that the revised procedure adversely affected the fire safety in the affected area. This decision was based on the recognition that the ability of the fire watch was not limited to identifying fires in a timely manner, but also on mitigation actions that an established fire watch could take in the event of fires. These actions could include the ability to close doors limiting fire exposure to adjacent areas, and providing more timely capability of detecting fires in certain cases. The inspectors concluded that the lack of a posted fire watch could adversely affect the ability to achieve and maintain safe shutdown in the event of a severe fire in the affected area (Ref. IR 05000482/200905).

Removal of Fire Protection Features without Compensatory Measures

In October 2002, NRC Region II inspectors found that a licensee made changes to the approved FPP which decreased the effectiveness of the program without prior Commission approval (Ref.: IR 05000260/2003007 and 05000296/2003007). Specifically, the licensee inappropriately used the License Condition Impact Evaluation (LCIE) change process to revise the FPP to allow the removal of fire suppression systems and/or fire-rated barrier assemblies installed to satisfy the separation and suppression requirements of 10 CFR 50, Appendix R, Sections III.G.2 and III.G.3, from service without implementing compensatory measures (i.e., fire watches). Before making this change, the licensee's NRC-approved FPP required that either a continuous or a one-hour compensatory fire watch patrol (with backup suppression equipment) be posted whenever a required fire suppression system and/or fire rated barrier assembly was inoperable. The LCIE concluded that the assignment and presence of fire watch personnel for the purpose of detecting and reporting fires with operable fire detection equipment was unnecessary and provided minimal additional fire protection safety margins. However, the evaluation did not include a technical basis for these conclusions.

In this case, the inspectors concluded that the licensee inappropriately used the fire protection program change process to revise the FPP to permit removing fire suppression systems and/or fire rated barrier assemblies from service without enhancing the other defense-in-depth elements as a compensatory measure. The change adversely affected the ability to achieve and maintain safe shutdown in the event of a fire, in that the licensee went from full compliance with the fire protection safe shutdown system separation and suppression criteria to less than

30

full compliance without implementing temporary measures to compensate for weakness in this defense-in-depth element. This was contrary to the safety objectives of the FPP and constituted a change from the approved program that required NRC approval prior to implementation.

Use of Operator Rounds to Compensate for Potential Circuit Vulnerabilities

The 1975 fire at Browns Ferry revealed the inadequacy of electrical isolation and physical separation practices in effect at that time. Failing to adequately identify circuits required to achieve and maintain safe shutdown and protect them from the effects of fire damage, could result in;

1) redundant trains of systems needed to achieve and maintain safe shutdown being damaged by fire; and,

2) the ability to safely shutdown the plant and maintain it in a safe shutdown condition being impaired.

In the mid-1990s, the NRC staff identified a regulatory concern about the potential impact of multiple spurious operations (MSO) that may occur as a result of fire damage to unprotected cables and circuits. These circuit analysis non compliances were judged to stem, in part, from a misunderstanding of the regulatory requirements and inappropriate reliance on a 1992 recommendation from the Nuclear Management and Resources Council (NUMARC now NEI) which was not endorsed by the NRC, and may have caused confusion during licensee evaluations (Ref. NUREG/BR-0195 Rev 2).

Based on the apparent widespread misunderstanding of regulatory requirements, enforcement discretion was deemed appropriate until the circuit analysis issue was resolved. On March 2, 1998, the NRC Office of Enforcement (OE) issued Enforcement Guidance Memorandum (EGM) 98-002 to temporarily defer formal enforcement actions to allow the industry time to develop positions that the NRC could endorse. During this time, licensees were to implement reasonable compensatory actions for identified circuit vulnerabilities. This initial guidance revised in July 1999 and February 2000.

On September 11, 2006, the staff proposed that the Commission issue a generic letter to clarify the fire-induced circuit failures. However, the staff's proposal was not approved by the Commission. Specifically, in SRM-SECY-06-0196, the Commission:

1) disapproved the proposed generic letter,

2) directed the staff to develop a clearly defined method of compliance to resolve fire-induced circuit failures for licensees who choose not to utilize the risk-informed approach in 10 CFR 50.48(c), and,

3) directed the staff to engage industry stakeholders to discuss the clarification of regulatory expectations to ensure a common understanding of the path to closure for this issue.

Subsequently, on June 30, 2008, the staff published SECY-08-0093, "Resolution of Issues Related to Fire Induced Circuit Failures." This SECY proposed a technical path forward to resolve multiple fire-induced circuit faults, including changes to enforcement guidance. On September 3, 2008, the Commission published SRM-SECY-08-0093 approving the staff's changes to the enforcement discretion for such circuit faults. Accordingly, on May 14, 2009, the NRC issued Enforcement Guidance Memorandum (EGM) 99-02 (Ref. Enforcement Guidance Memo 09-002) to provide a period of enforcement discretion for analyzing the effects of multiple fire-induced circuit faults. Specifically, EGM 99-02 permitted licensees to have until November 2, 2012, (3 years from the date of the issuance of Regulatory Guide (RG) 1.189, Revision 2, "Fire Protection for Nuclear Power Plants"[16]) to complete the corrective actions associated with the noncompliant multiple fire-induced circuit faults. Under EGM 99-02, licensees were expected to complete the following activities by November 2, 2012:

- identify noncompliance related to multiple fire-induced circuit faults

- implement adequate compensatory measures for these noncompliance,

- place the noncompliance in the corrective-action program, and,

- complete all corrective actions or submit a request for exemption or license amendment request for NRC review (Ref. Memorandum to NRC Region Administrators, 2010).

On May 13, 2010, the NRC staff held a public meeting with industry stakeholders to discuss, in part, industry's approach to compensatory measures for the multiple spurious-operation (MSO) non-compliances described in EGM 09-002. During the meeting, industry representatives described the use of existing, once-per-shift, operator rounds as a generic compensatory measure to satisfy EGM 09-002 requirements. In a subsequent, June 9, 2010, meeting summary (Ref. NRC Public Meeting Summary, ML101590181), the staff stated that sole reliance existing operator plant rounds that take place once per shift (i.e., every 8 or 12 hours depending on the shift's duration) is not an appropriate compensatory measure. The staff stated that MSOs that could affect safe shutdown are considered equivalent to a lack of a fire barrier and licensees should implement appropriate compensatory measures in accordance with their approved fire protection program.

During a subsequent meeting between the NRC Fire Protection Steering Committee and industry stakeholders on August 23, 2010 (Ref. NRC Public Meeting Summary, ML101590181), the staff again discussed the industry's use of enhanced operator rounds as a compensatory measure for MSO issues, and reiterated its position that using this approach is considered a noncompliance and licensees should implement compensatory measures consistent with the approved fire protection program.

4.2.3 Alternate Compensatory Measures

In certain instances, an approach different from the one specified in the approved FPP may be needed to provide an effective method of compensation. For example, properly analyzed Operator Manual Actions[17] (OMA) may be demonstrated to provide a more

---

[16] RG 1.189, Revision 2, issuance date was November 2, 2009 (74 FR 56673)

[17] In this context, "*properly analyzed*" means that the OMAs are appropriately justified by an evaluation. Criteria guidance acceptable to the staff for performing this evaluation is provided in Triennial Fire Protection Inspection Procedure IP 71111.05T.

effective compensatory measure for a degraded fire barrier than the hourly fire watch typically specified in an approved FPP. Similarly, a licensee may prefer to use new technologies, such those described in Appendix B, in lieu of or in conjunction with the measures specified in the approved FPP. For plants that have adopted the GL 86-10 Standard License Condition, the licensee may modify its FPP to allow their use of measures other than those specifically defined in the approved fire protection program, provided they would not adversely affect the ability to achieve and maintain safe shutdown in the event of a fire. This criterion is typically satisfied if the licensee can demonstrate that the alternate measure adequately offsets the degradation in defense-in-depth created by the inoperable, degraded, or nonconforming condition. In general, the evaluation should incorporate risk insights regarding:

- the location, quantity, and type of combustible material in the fire area,

- the presence of ignition sources and their likelihood of occurrence,

- the automatic fire suppression and fire detection capability in the fire area,

- the manual fire suppression capability in the fire area, and

- the human error probability where applicable.

The NRC staff has provided detailed information on the proper method for changing the approved fire protection program (FPP) to use an alternate compensatory in RIS 2005-07, "Compensatory Measures To Satisfy The Fire Protection Program Requirements" (Ref. RIS 2005-07).

4.3   Use of New Technologies

Many compensatory measures remain virtually unchanged from those that were put in place over 30 years ago in response to the Browns Ferry fire. In certain instances, these traditional measures may not be practical to implement, or may not provide an appropriate level of compensation for the degradation in defense-in-depth caused by the impairment. For example, the compensatory measure typically specified for a degraded fire barrier is an hourly fire watch. If, due to the specific hazards in the area, it is determined that sole reliance on a fire watch would not be sufficient to assure the ability to achieve and maintain safe shutdown conditions in the event of fire, a licensee would be expected to implement an alternate measure or combination of measures. Depending on the specific circumstances, such alternate compensatory measures may only involve additional enhancements to the compensatory measure specified in the FPP, such as operator briefings and interim shutdown strategies; or they may be more comprehensive such as installing temporary seals in fire barrier penetrations. In recent years, advances in fire technology have become available that, depending on the plant-specific conditions, may offer an effective alternate to or supplement for compensatory measures specified in the approved FPP. Several examples of new technologies that could potentially be used are illustrated in Appendix B.

5. EXAMPLE ASSESSMENT

As discussed previously, each operating plant has an approved FPP that integrates plant design features, personnel and administrative controls necessary to achieve an appropriate balance between each of the following elements of the defense-in-depth safety concept:

    (1) prevent fires from starting;
    (2) rapidly detect, control, and extinguish those fires that do occur; and
    (3) protect structures, systems, and components important to safety so that a fire that is not promptly extinguished will not prevent the safe shutdown of the plant.

The multiple layers of protection that are established by the defense-in-depth concept, offer reasonable assurance that deficiencies in one layer will not present an undue risk to public health and safety. Inherent in the DID safety concept is the need to ensure an appropriate level of fire safety is maintained when a fire protection feature is impaired or disabled. To achieve this objective, the FPP is designed to ensure appropriate compensatory measures are implemented whenever FPP features are degraded or impaired. The selection of a specific type of compensatory measure for a given impairment should be based on its ability to appropriately offset the degradation in DID created by the inoperable, degraded, or nonconforming condition.

Compensatory measures specified in a plant's approved FPP, are designed to provide an appropriate level of compensation for common impairments, such as those identified in Table 4-1 above. However, in certain unique cases, an appropriate compensatory measure may not be specified in the FPP. NRC Information Notice (IN) 97-48, "Inadequate or Inappropriate Interim Fire Protection Compensatory Measures," (Ref. US NRC, IN 97-48) notes that when unique cases are identified, appropriate compensatory measures should be determined by the licensee and tailored to the particular circumstances on a case-by-case basis. The selected measures should be consistent with the plant's license basis and demonstrated to be appropriate for the equipment or issue for which they were meant to compensate.

The hypothetical example provided in the following paragraphs is intended to illustrate how methods other than those specified in a plant's approved may be used to provide an effective alternative when unique circumstances are encountered.

5.1 Scenario

As illustrated in Figure 5-1, Fire Area SW-1A primarily contains equipment and cables associated with Shutdown Train A (shown in green). Thus, in the event of fire in this area, Train B (shown in red) systems and equipment would be relied on to achieve and maintain safe shutdown conditions. With the exception of control cables associated with a Train B motor-operated valve (MOV), one train of systems necessary to achieve hot shutdown is independent of the fire area. The cable trays containing this cable are protected throughout the area by an electrical raceway fire barrier system (i.e., fire wrap) having a 1-hour rating. In addition, the fire area is provided with area-wide ionization detectors and automatic sprinklers. However, consistent with the plant's fire protection licensing basis, sprinkler coverage limited a portion of the fire area that contains a high concentration of stacked cable trays. As shown in Figure 5-1, the automatic suppression system protects the Train B cable tray with the exception of a small segment that is routed near the 480V MCCs.

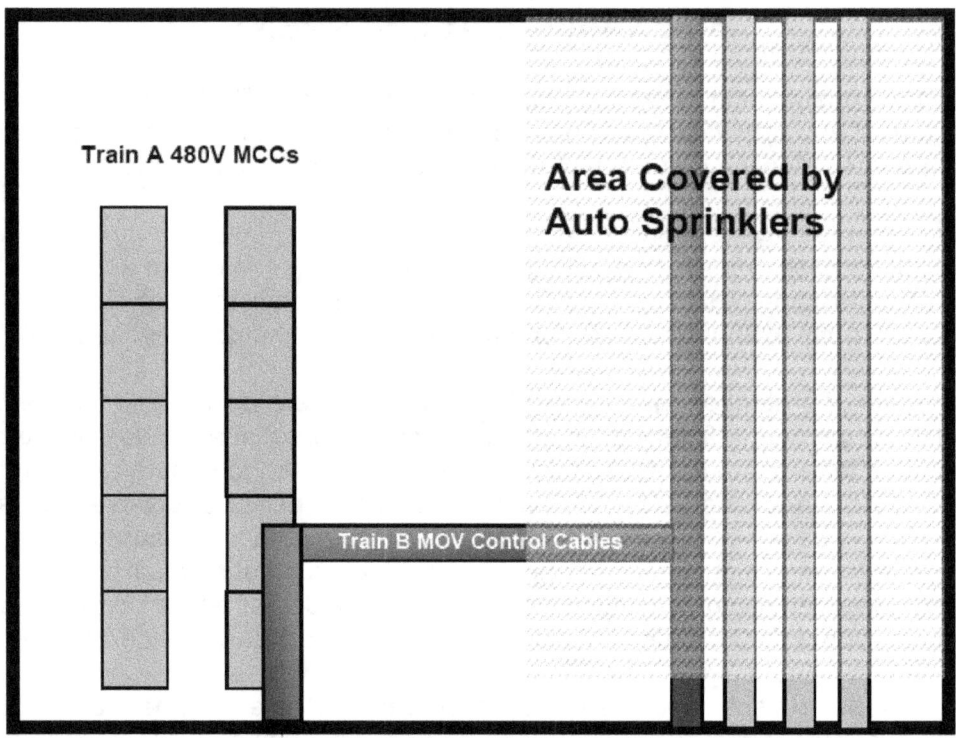

Figure 5-1. Fire Area SW-1A

Recently, the fire barrier covering the portion of the cable tray that is routed near the MCCs has been found have several significant through-cracks along its length. While not equivalent to a missing fire barrier, the degradation is considered to be sufficient to reduce the amount of time available for the fire brigade to manually extinguish a fire prior to cable damage. The specific "target" cable of concern is a multi-conductor PVC-insulated, control cable which transfers "open" and "close," position signals between the control room and a normally open MOV located in the suction line of a required pump. A fire-induced short circuit between two conductors in this cable would have the same effect as placing the MOV control switch to the "close" position. This would be unacceptable since this MOV must remain open to prevent pump damage due to loss of suction.

Because the fire area is provided with area-wide detection, the compensatory measure specified in the approved FPP is an hourly, roving, fire watch. However, due to the unique circumstances in this case (degraded protection of SSD capability), it is necessary to determine if the compensatory measure specified in the FPP (hourly fire watch) would provide a sufficiently effective means of compensation.

Scenario Summary

Degraded / Impaired FPP Feature:	Electrical Raceway Fire Barrier System (ERFBS)
DID Element Impacted:	3 (protect SSCs important to safety so that a fire that is not promptly extinguished will not prevent the safe shutdown of the plant)
FPP Compensatory Measure:	Hourly Fire Watch

5.2   Assessment

For this hypothetical scenario, the affected DID element is Element 3, (protect SSCs important to safety so that a fire that is not promptly extinguished will not prevent the safe shutdown of the plant). Therefore, the first step of the process is to determine the overall effectiveness of DID elements 1 and 2. For this example the following ranking scale will be used:

H = Highly Degraded,
M = Moderately Degraded, or
L = Low or no degradation – (Normal Operating State)

5.2.1   Evaluation of Defense-in-Depth Element 1 – Fire Prevention

The first element of DID is directed at preventing fires from starting. This is typically accomplished by controlling transient combustibles and ignition sources and preventing, to the extent practical, in situ ignition sources and combustible materials from causing self-sustaining fires. Examples of fire prevention activities include:

- Storing, handling, and using of flammable materials.
- Storage, control, and use of combustible materials relative to locations of flammable materials and ignition sources.
- Use and control of open flames and other ignition sources
- Housekeeping

Evaluation

Transient and fixed combustibles and ignition sources are found to be well controlled by established procedures in accordance with the plant's approved fire protection program. Thus, the Fire Prevention element of DID (DID Element 1) is judged to have a low degradation (Normal Operating State).

Fire Prevention Feature	Degradation Ranking
Transient Combustible Controls:	L
Transient Ignition Source Controls:	L

5.2.2   Evaluation of Defense-in-Depth Element 2 – Detection, Suppression, and Mitigation

The second element of DID is directed at rapidly detecting, controlling, and extinguishing fires that may occur in spite of the fire prevention efforts. This DID element is provided various types of fire protection features, such as: Heat and smoke detectors; fire alarm systems; Halon 1301, carbon dioxide (CO_2), and dry chemical fire suppression systems; automatic sprinkler, foam, and water spray systems; portable fire extinguishers; hose stations, fire hydrants, and water supply systems; and the fire brigades.

For the specific scenario under evaluation, DID element 2 is provided by the following systems:

- area-wide ionization detection system
- partial area automatic fire suppression
- manual suppression by fire brigade

Evaluation

Consistent with license basis requirements, area wide, ceiling mounted, spot-type detectors are installed and maintained to meet the NFPA code of record. In addition, partial sprinkler coverage is provided where a high concentration of cable trays are located. However, since the automatic suppression system does not provide coverage for the cable tray where the degraded fire barrier is located, it is judged to be moderately degraded.

As discussed in NRC Inspection Manual Chapter 609, Appendix F, (Ref. US NRC, IMC 609 App F Att. 3), cables protected by a fire barrier system with a 1-hour fire rating are subject to fire-induced failure if the fire barrier is degraded. Thus, a fire in FA SW-1A is expected to damage the cable of concern. Once the cable is damaged by fire, the type of fault needed to cause the MOV to spuriously change position (i.e., a conductor-to-conductor short in a multi-conductor cable) is expected to occur fairly quickly, prior to suppression by the plant fire brigade. Therefore the effectiveness of the fire brigade in preventing fire damage is also judged to be moderately degraded. Thus, although the area wide detection system will provide a prompt notification of fire, the impaired suppression capability causes DID Element 2 to be moderately degraded (M).

Fire Mitigation Feature	Degradation Ranking
Detection:	L – Normal Operating State
Automatic Sprinkler:	Moderately degraded with respect to the specific impairment
Manual Suppression:	Moderately degraded – with respect to the fire brigade's ability to extinguish a fire prior to cable damage.

38

5.2.3 Evaluation of Defense-in-Depth Element 3 – Protecting SSCs Important to Safety

The purpose of the third and final element of DID is to ensure that in the unlikely event that both DID elements 1 and 2 were to fail, the capability to achieve and maintain safe shutdown conditions would still be assured. Thus, this element of DID includes plant systems and design features that prevent fire from significantly damaging the SSCs required to achieve and maintain safe shutdown conditions. Examples include:

- Electrical Raceway Fire Barrier Systems (e.g., fire-resistant cable tray and conduit wraps)
- Fire Barriers (e.g., fire rated walls, floors, and ceilings)
- Electrical Isolation Devices (e.g., Isolation / Transfer Switches)
- Remote / Alternate Shutdown Equipment (e.g., Alternate Shutdown Panels)

In this example scenario, DID element 3 is primarily provided by the fire barrier wrap. However, the effectiveness of the barrier is degraded by several significant through-cracks. As discussed above for DID Element 2, cables protected by a fire barrier system with a 1-hour fire rating are considered to be subject to fire-induced failure if the fire barrier is degraded. Therefore, the target cable of concern can be expected to experience fire-induced faults as a result of fire in Fire Area SW-1A. Since fire damage that causes a short between two conductors of the target cable is sufficient to cause the MOV to move to an undesired (closed) position, spurious operation could occur prior to extinguishment by the plant fire brigade. Because pump damage on loss of suction is expected to occur fairly quickly, operators would not have sufficient time to restore operability of the affected pump.

Summary of DID Element 3

Due to the degraded ERFBS, a fire in Fire Are SW-1A would damage the target cable and, thereby, cause a normally open MOV to spuriously close. Due to its location in the flow path, unintended closing of this MOV would result in the loss of a makeup pump that is relied on to accomplish a hot shutdown function.

Capability to prevent cable damage (given a fire):	H
Capability to prevent spurious MOV actuation, given cable damage:	H
Capability of operators to recover	H

DID element 3 is judged to be Highly degraded.

5.3 Evaluation of Compensatory Measures

The licensee's FPP specifies that an hourly roving fire watch be implemented whenever an ERFBS is impaired and there is an operable detection system. However sole reliance on a fire watch as a compensatory measure for a safe shutdown function that is not adequately protected against fire damage may not be appropriate. As illustrated in this hypothetical scenario, the shutdown capability could be significantly affected if a fire were to occur during the time that the hourly fire watch is not in the area. To provide a more effective compensatory measure, the licensee has decided to enhance the level of protection provided by the hourly fire watch through the addition of advanced detection technologies capable of detecting fire in the incipient stages and operator guidance on actions necessary to prevent pump damage in the unlikely event a fire were to occur.

6. REFERENCES

1. Balaras, C.A.; "Infrared Thermography for Building Diagnostics, Energy and Buildings"; Volume 34, Issue 2, February 2002, Pages 171-183.

2. Bajwa, C.S.; West, K.S.; "Fire Barrier Penetrations Seals in Nuclear Power Plants"; U.S. NRC NUREG-1552; July 1996.

3. Brazzell, D.; "The Effects of High Air Velocity and Complex Airflow Patterns on Smoke Detector Performance" 2009.

4. Code of Federal Regulations (CFR), Title 10 Energy, Part 50, Section 50.36; "Technical Specifications"; U.S. Government Printing Office; Washington DC.

5. Code of Federal Regulations (CFR), Title 10, Energy, Part 54, Section 54.3; "Requirements For Renewal Of Operating Licenses For Nuclear Power Plants, Definitions"; U.S. Government Printing Office; Washington DC.

6. Cote, E.; "Fire Protection Handbook"; National Fire Protection Association; 17th Edition.

7. Custer, R., and Bright, R.; Technical Note 839: "Fire Detection: The State of the Art"; National Bureau of Standards; 1974.

8. Electric Power Research Institute (EPRI); TR 1013211: "Optimization of Fire Protection Impairments at Nuclear Power Plants," Final Report, August 2006.

9. Federal Register; Vol. 60, p. 42622 (60 FR 42622); "Use of Probabilistic Risk Assessment Methods in Nuclear Activities: Final Policy Statement".

10. Federal Register, Vol. 61, No. 70, p. 16011 (61 FR 16011); "All Licensees of Reactors with Installed Thermo-Lag Fire Barrier Material; Issuance of Director's Decision Under 10CFR2.206"; Wednesday, April 10, 1996.

11. Federal Register; Vol. 69, No. 115, p. 33536 (69 FR 33536); "Voluntary Fire Protection Requirements for Light Water Reactors; Adoption of NFPA 805 as a Risk-Informed, Performance-Based Alternative"; June 16, 2004.

12. Federal Register; Vol. 65, p. 38182 (65 FR 38182); "Elimination of the Requirement for Noncombustible Fire Barrier Penetration Seal Materials and Other Minor Changes"; June 20, 2000.

13. Gottuk, D.T.; Mckenna, L.A.; "Spot and Aspirated Laser Smoke Detection in Telecommunications Facilities"; Fire Technology 38, pages 147-178; 2002.

14. Kolaczkowski, A.; Forester, J.; Gallucci, R.; Klein, A.; Bongarra, J.; Qualls, P.; Barbadora, P.; Lois, E.; "Demonstrating the Feasibility and Reliability of Operator Manual Actions in Response to Fire"; US NRC NUREG-1852; October 2007.

15. Lewis, J.; "The 1975 Browns Ferry Fire Event"; Presentation; 2009 NEI Fire Protection Information Forum; ML092590396.

16. Liu, Z.G.; Kim, A.K.; "A Review of water mist fire suppression technology: Part II – Application Studies"; Journal of Fire Protection Engineering; Feb. 2001.

17. Mawhinney, J. R.; "A Review of Water Mist Fire Suppression Research and Development"; 1996; Fire Technology First Quarter 1997.

18. National Fire Protection Association (NFPA); Glossary of Terms; 2008 Edition.

19. National Fire Protection Association (NFPA); Standard 51B: "Standard for Fire Prevention during Welding, Cutting, and Other Hot Work"; 2009 Edition.

20. Nuclear Energy Institute (NEI) 04-02: "Guidance for Implementing a Risk-Informed, Performance-based Fire Protection Program under 10 CFR 50.48(c)"; Revision 2; April 2008.

21. U.S. NRC; Bulletin (BL) 92-01: "Failure of Thermo-Lag 330 Fire Barrier System to Maintain Cabling In Wide Cable Trays and Small Conduits Free From Fire Damage"; June 24, 1992.

22. U.S. NRC; Enforcement Action 96-131-Palisades (Consumer Power Company); "Notice of Violation and Proposed Imposition of Civil Penalty - $50,000 (NRC Inspection Report No. 50-255/96004 DRS)"; August 13, 1996.

23. U.S. NRC; Enforcement Guidance Memorandum (EGM) 09-002: "Enforcement Discretion for Fire Induced Circuit Faults"; May 14, 2009; ML090300446.

24. U.S. NRC, Frequently Asked Questions (FAQ) 08-0046: *Incipient Fire Detection Systems*; 2008; ML093220426

25. U.S. NRC; Generic Letter (GL) 80-100: "To All Power Reactor Licensees With Plants Licensed Prior To January 1, 1979," November 24, 1980.

26. U.S. NRC; Generic Letter GL 86-10: "Implementation of Fire Protection Requirements," April 24, 1986.

27. U.S. NRC; Generic Letter (GL) 88-12: "Removal of Fire Protection Requirements from Technical Specifications," August 2, 1988.

28. U.S. NRC; Information Notice (IN) 97-48: "Inadequate or Inappropriate Interim Fire Protection Compensatory Measures," July 9, 1997.

29. U.S. NRC; Inspection Manual Chapter (IMC) 609, Appendix F: "Fire Protection Significance Determination Process"; February 28, 2005.

30. U.S. NRC; Inspection Manual Chapter (IMC) 609, Appendix F, Attachment 3: "Guidance for Identifying Fire Growth and Damage Scenarios"; February 28, 2005.

31. U.S. NRC; Inspection Procedure (IP) 71111.15; "Operability Determinations and Functionality Assessments"; April 5, 2011; ML110030073.

32. U.S. NRC; Inspection Report (IR) 50-255/96-004; Palisades Nuclear Generating Plant; Accession Number 9605290105.

33. U.S. NRC; Inspection Report (IR) 50-458/97-201, "Fire Protection Functional Inspection"; River Bend Generating Station Unit 1; Accession Number 9803300071.

34. U.S. NRC; Inspection Report (IR) 05000260/2003007 and 05000296/2003007; "Browns Ferry Nuclear Plant - NRC Triennial Fire Protection"; Browns Ferry Nuclear Plant; November 17, 2003; ML033220576.

35. U.S. NRC; Inspection Report (IR) 05000482/2009005 and Notice of Violations; Wolf Creek Generating Station; February 11, 2010; ML100430713.

36. U.S. NRC Memorandum; "Recommended Fire Protection Policy and Program Actions," from the Fire Protection Policy Steering Committee To W.J. Dircks (Executive Director of Operations); October 26, 1984; ML031210015.

37. U.S. NRC Memorandum; "Application of the Enforcement Policy and Enforcement Guidance Memoranda on Fire Protection"; from R. P. Zimmerman (Director of the Office of Enforcement) to the Regional Administrators; April 16, 2010; ML101050187.

38. U.S. NRC Memorandum; "Completion of Review of Past Regulatory Instabilities Related to Nuclear Power Plant Fire Protection"; from A. Klein (NRR/Fire Protection Branch) to M.A. Cunningham (Director of NRR/Division of Risk Assessment); July 1, 2009; ML091690226.

39. U.S. NRC; NUREG 0050; "Recommendations Related to the Browns Ferry Fire: Report by Special Review Group"; February 1976.

40. U.S. NRC; NUREG-0103: "Standard Technical Specifications for Babcock and Wilcox Pressurized Water Reactors"; Revision 0, 1976; Revision 1, 1977; Revision 2, 1978; Revision 3, 1979; Revision 4, 1980. (Note: Later superseded by NUREG-1430)

41. U.S. NRC; NUREG-0123: "Standard Technical Specifications for General Electric Boiling Water Reactors"; Revision 0, 1976; Revision 1, 1978; Revision 2; 1979; Revision 3, 1980. (Note: Later superseded by NUREG-1433 [BWR/4] and NUREG-1434 [BWR/5])

42. U.S. NRC; NUREG/BR-0195: "Office of Enforcement Guidance Documents; Appendix A, Enforcement Guidance Memoranda"; Rev. 2.

43. U.S. NRC; NUREG-0212: "Standard Technical Specifications for Combustion Engineering Pressurized Water Reactors"; Revision 0, 1977; Revision 1, 1979; Revision 2, 1980; Revision 3, Draft, 1981. (Note: Later superseded by NUREG-1432)

44. U.S. NRC; NUREG-0298; Vol. 1; "Fire Protection Action Plan: Status Summary Report," February 1978.

45. U.S. NRC; NUREG-0452 "Standard Technical Specifications for Westinghouse Pressurized Water Reactors"; Revision 0, 1976; Revision 1, 1978; Revision 2, 1979;

Revision 3, 1980; Revision 4 1981; Revision 5, 1987. (Note: Later superseded by NUREG-1431)

46. U.S. NRC; NUREG-0800; "Standard Review Plan for the Review of Safety Analysis Reports for Nuclear Power Plants; Chapter 9.5.1: Fire Protection Program"; Rev. 4; October 2003.

47. U.S. NRC: Public Meeting Summary; "Summary of the May 13, 2010 Public Meeting to Discuss Fire Protection Topics Related to Circuits"; June 9, 2010; ML101590181.

48. U.S. NRC; Public Meeting Summary; "Public Meeting Summary of the August 23, 2010, Fire Protection Steering Committee"; September 15, 2010; ML102460061.

49. U.S. NRC; Regulatory Guide (RG) 1.189: "Fire Protection for Nuclear Power Plants," Rev. 2, October 2009.

50. U.S. NRC; Regulatory Guide (RG) 1.205: "Risk-Informed, Performance-Based Fire Protection for Existing Light-Water Nuclear Power Plants"; Revision 1; December 2009.

51. U.S. NRC; Regulatory Information Summary (RIS) 2005-07: "Compensatory Measures To Satisfy The Fire Protection Program Requirements," April 19, 2005.

52. U.S. NRC; Regulatory Issue Summary (RIS) 2005-20: "Revision to NRC Inspection Manual Part 9900 Technical Guidance, "Operability Determinations & Functionality Assessments for Resolution of Degraded or Nonconforming Conditions Adverse to Quality or Safety""; Revision 1; April 16, 2008.

53. U.S. NRC; SECY-96-146; "Technical Assessment of Fire Barrier Penetration Seals in Nuclear Power Plants"; July 1, 1996.

54. U.S. NRC; SECY-98-0058; "Development of a Risk-Informed, Performance-Based Regulation for Fire Protection at Nuclear Power Plants"; March 26, 1998.

55. U.S. NRC; SECY-00-0009; "Rulemaking Plan, Reactor Fire Protection Risk-Informed, Performance-Based Rulemaking"; January 13, 2000.

56. U.S. NRC; SECY 08-0171, "Plan for Stabilizing Fire Protection Regulatory Infrastructure"; dated November 5, 2008.

57. U.S. NRC; Staff Requirements Memorandum (SRM) M080717: "Briefing on Fire Protection Issues"; July 29, 2008.

58. U.S. NRC; Task Interface Agreement (TIA) (96TIA008); "NRR Response to Region IV: Evaluation of Definition of Continuous Fire Watch"; August 17, 1998; ML0124000480.

59. Salley, M.H.; Iqbal, N.; "Fire Dynamics Tools, Quantitative Fire Hazard Analysis Methods for the U.S. Nuclear Regulatory Commission Fire Protection Inspection Program"; US NRC NUREG-1805: Final Report; December 2004.

60. Shearon Harris Nuclear Power Plant, Unit 1; "Issuance Of Amendment Regarding Adoption Of National Fire Protection Association Standard 805: Performance-Based

Standard For Fire Protection For Light Water Reactor Electric Generating Plants"; (TAC NO. MD8807); June 28, 2010.

61. Society of Fire Protection Engineers; Technology Report 77-2: "The Browns Ferry Nuclear Plant Fire."

7. GLOSSARY

Below are key terms or phrases whose definitions and associated context, for purposes of this document, are as shown.

Administrative Controls	Policies, procedures, and other elements that relate to the FPP. Administrative controls include but are not limited to inspection, testing, and maintenance of fire protection systems and features; compensatory measures for fire protection impairments; review of the impact of plant modifications on the FPP; fire prevention activities; fire protection staffing; control of combustible/flammable materials; and control of ignition sources. (RG 1.189 R2)
Air Sampling Detector	A detector that consists of a piping or tubing distribution network that runs from the detector to the area(s) to be protected. An aspiration fan in the detector housing draws air from the protected area back to the detector through air sampling ports, piping, or tubing. At the detector, the air is analyzed for fire products. (NFPA 72 2007)
Alternate Compensatory Measure	Alternate compensatory measures are those which differ from measures identified in the approved fire protection program. For example, if the approved FPP identifies a fire watch as the only appropriate compensatory measure for a degraded fire barrier, the use of a video image smoke detection system in lieu of a fire watch would be considered an "alternate compensatory measure." Ref.: RIS 2005-07, "Compensatory Measures To Satisfy The Fire Protection Program Requirements."
Alternate Shutdown	The capability to shut down the reactor that is required when it is not feasible to provide the required protection for redundant safe-shutdown trains in one or more fire areas or where fire suppression activities, including inadvertent operation or rupture of a suppression system, could prevent safe shutdown. Appendix R to 10 CFR Part 50 allows an existing plant system to be rerouted, relocated, or modified (at the time the need for an alternative means of shutdown is identified but not during or after the fire) such that it can perform the required safe-shutdown function that the redundant system damaged by fire or damaged by suppression system discharge would normally perform. (RG 1.189 R2)
Approved	Tested and accepted for a specific purpose or application by a recognized testing laboratory or reviewed and specifically approved by the NRC in accordance with the appropriate regulatory process. (RG 1.189 R2)
Automatic	Self-acting, operating by its own mechanism when actuated by some monitored parameter such as a change in current, pressure, temperature, or mechanical configuration. (RG 1.189 R2)

Compensatory Actions

Actions taken if an impairment to a required system, feature, or component prevents that system, feature, or component from performing its intended function. These actions are a temporary alternative means of providing reasonable assurance that the necessary function will be compensated for during the impairment, or an act to mitigate the consequence of a fire. Compensatory measures include but are not limited to actions such as firewatches, administrative controls, temporary systems, and features of components. (NFPA 805) (NFPA Glossary 2008)

Configuration Management

Configuration management is an integrated management process used to ensure that the licensee maintains the plant's physical and functional characteristics in conformance with its design and licensing basis; that operating, training, modification, and maintenance processes are consistent with the conditions prescribed by the design and current licensing basis; and that the licensee operates and maintains the plant within these conditions (NUREG-1913)

Current License Basis

The current licensing basis (CLB) is the set of NRC requirements applicable to a specific plant and a licensee's written commitments for ensuring compliance with and operation within applicable NRC requirements and the plant-specific design basis (including all modifications and additions to such commitments over the life of the license) that are docketed and in effect. The CLB includes:

- NRC regulations contained in 10 CFR Parts 2, 19, 20, 21, 26, 30, 40, 50, 51, 54, 55, 70, 72, 73, 100, and appendices thereto;
- Commission orders;
- License Conditions;
- Exemptions;
- Technical Specifications.
- Plant specific design-basis information defined in 10 CFR 50.2 as documented in the most recent FSAR, as required by 10 CFR 50.71
- Licensee's commitments remaining in effect that were made in docketed licensing correspondence such as licensee responses to NRC bulletins, generic letters, and enforcement actions,
- Licensee commitments documented in NRC safety evaluations or licensee event reports (LERs). (10 CFR 54.3(a))

Defense-in-Depth	Fire protection for nuclear power plants uses the concept of defense-in-depth to achieve the required degree of reactor safety. This concept entails the use of echelons of administrative controls, fire protection systems and features, and safe-shutdown capability to achieve the following objectives:

  • Prevent fires from starting.
  • Detect rapidly, control, and extinguish promptly those fires that do occur.
  • Protect SSCs important to safety, so that a fire that is not promptly extinguished by the fire suppression activities will not prevent the safe shutdown of the plant. (RG 1.189 R2)

  Defense-in-depth is an element of the NRC's Safety Philosophy that employs multiple layers or echelons of protection to prevent accidents or mitigate damage if a malfunction, accident, or naturally-caused event occurs at a nuclear facility. The defense-in-depth philosophy ensures that safety will not be wholly dependent on any single element of the design, construction, maintenance, or operation of a nuclear facility. As stated in NUREG 0050, "Recommendations Related to the Browns Ferry Fire." the goal is to provide a suitable balance of these multiple layers or echelons of protection. Increased strength, redundancy, performance, or reliability of one echelon can compensate in some measure for deficiencies in the others.

Degraded Condition	A condition of an SSC in which there has been any loss of quality or functional capability. (GL 91-18) A degraded condition is one in which the qualification of an SSC or its functional capability is reduced. Examples of degraded conditions are failures, malfunctions, deficiencies, deviations, and defective material and equipment. Examples of conditions that can reduce the capability of a system are aging, erosion, corrosion, improper operation, and maintenance. (NUREG 1913)
Design Basis	*Design bases* means that information which identifies the specific functions to be performed by a structure, system, or component of a facility, and the specific values or ranges of values chosen for controlling parameters as reference bounds for design. These values may be (1) restraints derived from generally accepted "state of the art" practices for achieving functional goals, or (2) requirements derived from analysis (based on calculation and/or experiments) of the effects of a postulated accident for which a structure, system, or component must meet its functional goals (10 CFR 50.2) Design basis is that body of plant-specific design bases information defined by 10 CFR 50.2 (GL 91-18).
Electrical Raceway Fire Barrier System (ERFBS)	Non-load-bearing partition type envelope system installed around electrical components and cabling that are rated by test laboratories in hours of fire resistance and are used to maintain safe-shutdown functions free of fire damage. (RG 1.189 R2)

Fire	A rapid oxidation process, which is a chemical reaction resulting in the evolution of light and heat in varying intensities. (NFPA Glossary 2008)
Fire Alarm System	A system or portion of a combination system that consists of components and circuits arranged to monitor and annunciate the status of fire alarm or supervisory signal-initiating devices and to initiate the appropriate response to those signals. (NFPA 1 2012)
Fire Area	The portion of a building or plant that is separated from other areas by rated fire barriers adequate for the fire hazard. (RG 1.189 R2)
Fire Barrier	Components of construction (walls, floors, and their supports), including beams, joists, columns, penetration seals or closures, fire doors, and fire dampers, that are rated by approving laboratories in hours of resistance to fire that are used to prevent the spread of fire. (RG 1.189 R2) A continuous membrane or a membrane with discontinuities created by protected openings with a specified fire protection rating, where such membrane is designed and constructed with a specified fire resistance rating to limit the spread of fire, that also restricts the movement of smoke. (NFPA 101 2003) In nuclear facilities, a continuous assembly designed and constructed to limit the spread of heat and fire and to restrict the movement of smoke. (NFPA 805)
Fire Brigade	A team of onsite plant personnel that is qualified and equipped to perform manual fire suppression activities. (RG 1.189 R2)
Fire Damage	The total damage to a building, structure, vehicle, natural vegetation cover, or outside property resulting from a fire and the act of controlling that fire. (NFPA Glossary 2008)
Fire Hazards	Conditions that involve the necessary elements to initiate and support combustion, including in situ or transient combustible materials, ignition sources (e.g., heat, sparks, open flames), and an oxygen environment. (RG 1.189 R2)
Fire Hazards Analysis	An analysis used to evaluate the capability of a nuclear power plant to perform safe-shutdown functions and minimize radioactive releases to the environment in the event of a fire. The analysis includes the following features: • identification of fixed and transient fire hazards • identification and evaluation of fire prevention and protection measures relative to the identified hazards • evaluation of the impact of fire in any plant area on the ability to safely shut down the reactor and maintain shutdown conditions, as well as to minimize and control the release of radioactive material (RG 1.189 R2) An analysis to evaluate potential fire hazards and appropriate fire protection systems and features used to mitigate the effects of fire in any plant location. (NFPA 805)

Fire Protection Feature	Administrative controls, emergency lighting, fire barriers, fire detection and suppression systems, fire brigade personnel, and other features provided for fire protection purposes. (RG 1.189 R2)

Administrative controls, fire barriers, means of egress, industrial fire brigade personnel, and other features provided for fire protection purposes. (NFPA 805) |
| Fire Protection Plan | 10 CFR 50.48(a) requires licensees to have a *fire protection plan* that meets General Design Criterion (GDC) 3 of Appendix A to 10 CFR Part 50. In accordance with 10 CFR 50.48, the *fire protection plan* must do the following:

- Describe the overall Fire Protection Program for the facility.
- Identify the various positions within the licensee's organization that are responsible for the program.
- State the authorities that are delegated to each of these positions to implement those responsibilities.
- Outline the plans for fire protection, fire detection and suppression, and limitation of fire damage.
- Describe the administrative controls and personnel requirements for fire protection and manual fire suppression activities.
- Describe the automatic and manually operated fire detection and suppression systems.
- Describe the means to limit fire damage to SSCs important to safety to ensure the ability to shut down the plant safely. |
Fire Protection Program	The integrated effort involving components, procedures, analyses, and personnel utilized in defining and carrying out all activities of fire protection. It includes system and facility design, fire prevention, fire detection, annunciation, confinement, suppression, administrative controls, fire brigade organization, inspection and maintenance, training, quality assurance, and testing. (RG 1.189 R2)
Fire Protection Program Attribute	*Fire Protection Program Attributes* are characteristics of the fundamental FPP elements and may vary based on the plant licensing basis (SECY-04-0050)
Fire Protection Program Elements	*Fire Protection Program Elements* are the fundamental features or components of the approved fire protection program (SECY-04-0050)
Fire Protection System	Fire detection, notification, and suppression systems designed, installed, and maintained in accordance with the applicable nationally recognized codes and standards endorsed by the NRC. (RG 1.189 R2)
Any fire alarm device or system or fire extinguishing device or system, or combination thereof, that is designed and installed for detecting, controlling, or extinguishing a fire or otherwise alerting occupants, or the fire department, or both, that a fire has occurred. (NFPA 1)
Fire detection, notification, and fire suppression systems designed, installed, and maintained in accordance with the applicable NFPA codes and standards. (NFPA 805)

Fire Scenario	A set of conditions that defines the development of fire, the spread of combustion products throughout a building or portion of a building, the reactions of people to fire, and the effects of combustion products. (NFPA 101 2012)
Fire Suppression	Control and extinguishing of fires (firefighting). Manual fire suppression is the use of hoses, portable extinguishers, or manually actuated fixed systems by plant personnel. Automatic fire suppression is the use of automatically actuated fixed systems such as water, Halon, or CO_2 systems. (RG 1.189 R2)
Fire Watch	Individuals responsible for providing additional (e.g., during hot work) or compensatory (e.g., for system impairments) coverage of plant activities or areas for the purposes of detecting fires or for identifying activities and conditions that present a potential fire hazard. The individuals should be trained in identifying conditions or activities that present potential fire hazards, as well as the use of fire extinguishers and the proper fire notification procedures. (RG 1.189 R2) The assignment of a person or persons to an area for the express purpose of notifying the fire department, the building occupants, or both of an emergency; preventing a fire from occurring; extinguishing small fires; or protecting the public from fire or life safety dangers. (NFPA 1 2012)
Fire Zones	Fire zones are subdivisions of fire areas. (RG 1.189 R2) (NUREG-1805) A subdivision of a fire area not necessarily bounded by fire-rated assemblies. Fire zone can also refer to the area subdivisions of a fire detection or suppression system, which provide alarm indications at the central alarm panel. (NFPA 805) Fire zones are subdivisions of fire areas defined in the context of the fire protection program. A fire zone is not necessarily bounded by fire barriers. Zone divisions are often defined based on the fire suppression and/or detection systems designed to combat particular types of fires. A fire zone may contain one or more rooms. A fire compartment may contain one or more fire zones. (NUREG/CR 6850)
Functional / Functionality	*Functionality* is an attribute of SSCs that is not controlled by TSs. An SSC is functional or has functionality when it is capable of performing its specified function, as set forth in the CLB. Functionality does not apply to specified safety functions, but does apply to the ability of non-TS SSCs to perform other specified functions that have a necessary support function. To be considered operable, a non-TS SSC does not have to be "fully qualified" in terms of its design and licensing bases as long as the licensee can demonstrate functionality. For example, if a fire protection SSC that was removed from the TS in accordance with GL 86-10 is determined to be nonfunctional, a functionality assessment should be performed to determine if it is capable of performing the function specified in the plant's current licensing basis. (RIS 2005-20)

Heat Detector	A fire detector that detects either abnormally high temperature or rate of temperature rise, or both. (NFPA 72, 2010)
Hot Work	Activities that involve the use of heat, sparks, or open flame such as cutting, welding, and grinding. (RG 1.189 R2)
Impairment	The degradation of a fire protection system or feature that adversely affects the ability of the system or feature to perform its intended function. (RG 1.189 R2) Impairment exists if a fire protection feature (both active and passive) is no longer capable of performing the function(s) specified in the current fire protection licensing basis.
Incipient Stage Fire	A fire which is in the initial or beginning stage and which can be controlled or extinguished by portable fire extinguishers or small hose systems without the need for protective clothing or breathing apparatus. (NFPA Glossary 2008)
In situ Combustibles	Combustible materials that constitute part of the construction, fabrication, or installation of plant SSCs and as such are fixed in place. (RG 1.189 R2) Combustible materials that are permanently located in a room or an area (e.g., cable insulation, lubricating oil in pumps). (NFPA 805)
Interim Compensatory Measures	Actions taken if an impairment to a required system, feature, or component prevents that system, feature, or component from performing its intended function. These actions are a temporary alternative means of providing reasonable assurance that the necessary function will be compensated for during the impairment, or an act to mitigate the consequence of a fire. Compensatory measures include but are not limited to actions such as fire watches, administrative controls, temporary systems, and features of components. (NFPA 805) (NFPA Glossary 2008). See Compensatory Measures
Limiting Condition for Operation	The section of Technical Specifications that identifies the lowest functional capability or performance level of equipment required for safe operation of the facility.
Mitigate	Perform an action that stops the progression of or reduces the severity of an unwanted condition. With respect to nuclear plant fire protection, mitigation generally refers to operator actions inside or outside the main control room to restore the capability to achieve and maintain safe shutdown where a fire has degraded that capability. (RG 1.189 R2) Activities taken to eliminate or reduce the degree of risk to life and property from hazards, either prior to or following a disaster/emergency. (NFPA Glossary 2003)

Noncombustible Material	(a) Material that, in the form in which it is used and under conditions anticipated, will not ignite, burn, support combustion, or release flammable vapors when subjected to fire or heat, or (b) material having a structural base of noncombustible material, with a surfacing not over 3 mm (c inch) thick that has a flame spread rating not higher than 50 when measured in accordance with ASTM E-84, "Standard Test Method for Surface Burning Characteristics of Building Materials" (RG 1.189 R2) A substance that will not ignite and burn when subjected to a fire. (NFPA Glossary 2008)
Nonconforming Condition	A condition of an SSC in which there is failure to meet requirements or licensee commitments (GL 91-18). Some examples of nonconforming conditions include the following: • There is failure to conform to one or more applicable codes or standards specified in the FSAR. • As-built equipment, or as-modified equipment, does not meet FSAR descriptions. • Operating experience or engineering reviews demonstrate a design inadequacy. • Documentation required by NRC requirements such as 10 CFR 50.49 is not available or deficient.
NRC	*NRC* means the Nuclear Regulatory Commission, the agency established by title II of the Energy Reorganization Act of 1974, as amended, comprising the members of the Commission and all offices, employees, and representatives authorized to act in any case or matter (see also *Commission*).
Onsite	As defined in the site specific UFSAR (usually refers to the owner-controlled area or real estate over which the licensee has the legal right to control access.).
Operable	A system, subsystem, train, component, or device is considered operable or have operability when it is capable of performing its specified safety function(s) This includes adequate performance of any support instrumentation, controls, electrical power, cooling or seal water, lubrication or auxiliary equipment. (NUREG 1913) **Note**: *Operable* applies to SSCs identified in the TS. Existing plant-specific TSs contain several variations on this basic definition. In all cases, a licensee's plant-specific TS definition of Operable/Operability governs. (IMC 9900, Operability Determination Process 9/26/05). *Functionality* applies to non-TS SSCs.
Operator Manual Action (OMA)	Actions performed by operators to manipulate components and equipment from outside the main control room not including "repairs." (RG 1.189 R2) Those actions performed by operators to manipulate components and equipment from outside the main control room to achieve and maintain postfire hot shutdown, but not including "repairs." Operator manual actions comprise an integrated set of actions needed to help ensure that hot shutdown can be accomplished, given that a fire has occurred in a particular plant area. (NUREG-1852)

Passive Fire Protection Feature	A fire protection feature that provides a means of controlling a fire without a distinct change of state. Fire barriers, cable tray fire wrap, and penetration seals are examples of passive fire protection features. Protection measures that prolong the fire resistance of an installation before an eventual fire occurrence from the effects of smoke, flames, and combustion gases. These can consist of insulation (fireproofing) of a structure, choice of noncombustible materials of construction, use of fire-resistant partitions, and compartmentation to resist the passage of fire. It includes coatings, claddings, or free-standing systems that provide thermal protection in the event of fire and that require no manual, mechanical, or other means of initiation, replenishment, or sustainment for their performance during a fire incident. (NUREG-1805)
Penetration	An opening for penetrations that pass through both sides of a vertical or horizontal fire resistance–rated assembly. (NFPA 5000 2012)
Performance-Based Approach	A performance-based approach relies upon measurable (or calculable) outcomes (i.e., performance results) to be met but provides more flexibility as to the means of meeting those outcomes. A performance-based approach is one that establishes performance and results as the primary basis for decision-making and incorporates the following attributes: (1) Measurable or calculable parameters exist to monitor the system, including facility performance; (2) Objective criteria to assess performance are established based on risk insights, deterministic analyses, and/ or performance history; (3) Plant operators have the flexibility to determine how to meet established performance criteria in ways that will encourage and reward improved outcomes; and (4) A framework exists in which the failure to meet a performance criteria, while undesirable, will not in and of itself constitute or result in an immediate safety concern. (NFPA 805)
Post-Fire Safe-Shutdown Analysis	A process or method of identifying and evaluating the capability of SSCs necessary to accomplish and maintain safe-shutdown conditions in the event of a fire. (RG 1.189 R2)
Post-Fire Safe-Shutdown System/Equipment	Systems and equipment that perform functions needed to achieve and maintain safe shutdown during and following a fire (regardless of whether the system or equipment is part of the success path for safe shutdown). This includes systems and equipment of which fire-induced spurious actuation could prevent safe shutdown. (RG 1.189 R2)
Pre-Fire (Incident) Plans	Documentation that describes the facility layout, access, contents, construction, hazards, hazardous materials, types and locations of fire protection systems, and other information important to the formulation and planning of emergency fire response. (RG 1.189 R2) A written document resulting from the gathering of general and detailed data to be used by responding personnel for determining the resources and actions necessary to mitigate anticipated emergencies at a specific facility. (NFPA Glossary 2008)

Quality Assurance	"Quality assurance" comprises all those planned and systematic actions necessary to provide adequate confidence that a structure, system, or component will perform satisfactorily in service. Quality assurance includes quality control, which comprises those quality assurance actions related to the physical characteristics of a material, structure, component, or system which provide a means to control the quality of the material, structure, component, or system to predetermined requirements. (10 CFR 50, Appendix B—Quality Assurance Criteria for Nuclear Power Plants and Fuel Reprocessing Plants) Attributes of a QA program include programs that preserve quality through procedures, recordkeeping, inspections, corrective actions, and audits. NUREG-1913)
Redundancy	Redundancy refers to the provision of multiple independent methods of equivalent capacity to perform a specified function.
Redundant train or system	One of two or more similar trains of equivalent capacity in the same system powered by separate electrical divisions or one of two or more separate systems that each perform the same post-fire safe-shutdown function as its design function. With respect to fire protection regulatory requirements and guidance, a redundant train or system is distinct from an alternative or dedicated shutdown train or system. (RG 1.189 R2)
Risk	The set of probabilities and consequences for all possible accident scenarios associated with a given plant or process. (NFPA 805)
Risk Informed Approach	A philosophy whereby risk insights are considered together with other factors to establish performance requirements that better focus attention on design and operational issues commensurate with their importance to public health and safety. (NFPA 805)
Safe-Shutdown Analysis	A process or method of identifying and evaluating the capability of SSCs necessary to accomplish and maintain safe-shutdown conditions in the event of a fire. (RG 1.189 R2)
Safety-Related Function	A safety-related function applies to the SSCs, procedures, and controls of a facility or process that must remain functional during and following design-basis events to ensure the integrity of the facility's reactor coolant pressure boundary, the facility's capability to shut down the reactor and maintain it in a safe shutdown condition, or the facility's capability to prevent or mitigate the consequences of accidents that could result in potential offsite exposure comparable to the guidelines in 10 CFR 50.34(a)(1), 10 CFR 50.67(b)(2), or 10 CFR 100.11, "Determination of Exclusion Area, Low Population Zone, and Population Center Distance." An example of a safety related function is a facility's capability to shut down a nuclear reactor and maintain it in a safe shutdown condition. (NUREG-1913)
Smoke Detector	A device that detects visible or invisible particles of combustion. (NFPA 72 2010)

Spot-Type Detector	A device in which the detecting element is concentrated at a particular location. Typical examples are bimetallic detectors, fusible alloy detectors, certain pneumatic rate-of-rise detectors, certain smoke detectors, and thermoelectric detectors. (NFPA 72 2010)
Standards (Code) of Record	The specific editions of the nationally recognized codes and standards accepted by the NRC that constitute the licensing and design basis for the plant. (see Code of Record)
Standard Technical Specifications	NRC staff guidance on model technical specifications for an operating license. (NRC Online Glossary) Standard Technical Specifications (STS) are published for each of the five reactor types as a NUREG-series publication. Plants are required to operate within these specifications. (NRC Technical Specifications web page)
Standpipe and Hose Systems	An arrangement of piping, valves, hose connections, and allied equipment installed in a building or structure, with the hose connections located in such a manner that water can be discharged in streams or spray patterns through attached hose and nozzles, for the purpose of extinguishing a fire, thereby protecting a building or structure and its contents in addition to protecting the occupants. (NFPA 14 2010) The provision of piping, riser pipes, valves, firewater hose connections, and associated devices for the purpose of providing or supplying firewater hose applications in a building or structure by the occupants or fire department personnel. (NUREG-1805)
Technical Specification (TS)	Part of an NRC license authorizing the operation of a nuclear production or utilization facility. A Technical Specification establishes requirements for items such as safety limits, limiting safety system settings, limiting control settings, limiting conditions for operation, surveillance requirements, design features, and administrative controls. (NRC Online Glossary)
Transient Combustibles	Combustible materials that are not fixed in place or not an integral part of an operating system or component. (RG 1.189 R2)

APPENDIX A: COMPARISON OF ACTIONS SPECIFIED IN STANDARDIZED TECHNICAL SPECIFICATIONS TO COMPENSATORY MEASURES IDENTIFIED IN APPROVED FIRE PROTECTION PROGRAMS

This section provides a comparison of compensatory measures specified in Historical Standardized Technical Specifications (prior to GL 86-10) to those specified in a sample of approved Fire Protection Programs.

Table A-1. Comparison of Technical Specification Compensatory Actions to Current FPP Measures				
Affected System or Feature	Impairment	Original STS Action[18]	Current FPP Actions [19]	
			Plant	Action
Fire Suppression Water Supply System	One fire pump and/or one water supply inoperable	Restore to operable status within 7 days	A	Restore the water supply system to operable station within 7 days or provide an alternate backup pump or supply
			B	Restore the inoperable equipment to OPERBLE status within 7 days.
			C	Restore the inoperable equipment to operable status within 7 days.
			D	Establish hourly fire watch in fire areas where design flow requirements for required spray/sprinkler systems are not met using the electric motor-driven pump(s). Provide alternate pumping capability, OR Provide backup pumping capability and maintain fire watch
			E	Restore the inoperable pump and/or water supply to operable status within 7 days, OR provide an alternate backup pump and/or water supply.
			F	Restore on additional pump to operable/functional status OR provide an alternate pump within 7 days.
			G	Restore pump to operable status or provide a backup fire pump in 14 Days
Fire Suppression Water Supply System	Both fire pumps and/or both water supplies inoperable	Establish backup fire suppression within 24 hours	A	Establish a backup fire protection water system within 24 hours OR within one hour initiate action to place the plant in Hot Shutdown within the following 6 hours, and Cold Shutdown within the subsequent 24 hours.

---

[18] Based on NRC/RES review of the Standard Technical Specifications. Fire Protection was added to the STS in Revision 1 of each of the NSSS Vendor STS (NUREG-0103, 0123, 0212 and 0452).

[19] As specified in available revisions of plant specific TRMs and / or FPP implementing procedures developed in response to GL 86-10

Affected System or Feature	Impairment	Original STS Action[18]	Current FPP Actions [19]	
			Plant	Action
			B	Establish a backup fire suppression water system within 24 hours, OR Be in at least Hot Shutdown within the next 12 hours and in Cold Shutdown within the following 24 hours.
			C	Establish a backup fire suppression water system within 24 hours or place the reactor in Hot Standby within the next six (6) hours and in Cold Shutdown within the following 30 hours.
			D	Provide a backup supply within 24 hours And Establish hourly fire watch in fire areas where design flow requirements for required / specified spray/sprinkler systems are not met
			E	Establish a backup fire suppression water system within 24 hours.
			F	Restore at least one water supply source OR establish an alternate fire water supply within 24 hours.
			G	Provide a backup fire pump within 24 hours. If completion times are not met Be in Mode 3 within 7 hours AND Be in Mode 4 in 13 hours AND Be in Mode 5 in 37 hours.
Spray and/or Sprinkler Systems	one or more of the required spray/sprinkler systems inoperable in areas containing redundant trains of shutdown equipment, Outside Containment	Continuous fire watch with back up fire suppression	A	Continuous fire watch with backup fire suppression equipment
			B	Continuous fire watch with backup fire suppression equipment
			C	Continuous fire watch with backup fire suppression equipment
			D	Continuous fire watch with backup fire suppression equipment
			E	Continuous fire watch with backup fire suppression equipment

Table A-1. Comparison of Technical Specification Compensatory Actions to Current FPP Measures

Table A-1. Comparison of Technical Specification Compensatory Actions to Current FPP Measures

Affected System or Feature	Impairment	Original STS Action[18]	Current FPP Actions [19]	
			Plant	Action
			F	Continuous fire watch with backup fire suppression equipment
			G	Continuous fire watch with backup fire suppression equipment
Fire Suppression Sprinkler / Spray Systems	One or more spray / sprinklers inoperable in areas NOT containing redundant trains of shutdown equipment (Outside Containment)	Hourly Fire Watch	A	Hourly fire watch
			B	Hourly fire watch
			C	Hourly fire watch
			D	Hourly fire watch
			E	Hourly fire watch
			F	Hourly fire watch
			G	Hourly fire watch
Fire Suppression CO2 Systems (TS Rev 2: Combines LP and HP Systems in the same section)	One or more of the required LP CO2 systems inoperable	Establish a continuous fire watch with back up fire suppression equipment for the unprotected areas within 1 hour STS Rev 2 changed this to fire watch with fire suppression equipment for those areas redundant systems or components and for other areas an hourly fire watch patrol.	A	(Non-essential): Issue a Fire Protection System Impairment (FPSI) permit. Evaluate the potential impact on the impairment and implement appropriate compensatory actions.
			B	Establish a continuous fire watch and backup fire suppression equipment for the unprotected re(s) within 1 hour AND restore the system to OPERABLE status within 14 days AND Place signs at the backup fire suppression equipment to identify the proper hose to be used.
			C	Not Applicable.
			D	Not Applicable.

Table A-1. Comparison of Technical Specification Compensatory Actions to Current FPP Measures

Affected System or Feature	Impairment	Original STS Action[18]	Current FPP Actions [19]	
			Plant	Action
			E	Establish a continuous fire watch with backup fire suppression equipment for those areas in which redundant systems or components could be damaged within 1 hour OR for those areas not having redundant systems or components establish a roving watch.
			F	Not Applicable.
			G	Not Applicable.
Fire Suppression Halon Systems	One or more Halon systems inoperable	With one or more of the required Halon systems inoperable, establish a continuous fire watch with back up fire suppression equipment for the unprotected areas within 1 hour STS Rev 2 changed this to fire watch with fire suppression equipment for those areas redundant systems or components and for other areas an hourly fire watch patrol.	A	Issue an Fire Protection System Impairment (FPSI) permit for the inoperable system. A fire tour is not required in the Min Control Room since it is continually manned.
			B	(DG Building Basement): Establish an hourly fire watch with backup fire suppression within one hour AND restore the system to OPERABLE status within 14 days
			C	Establish a continuous fire watch with backup fire suppression equipment in the area protected within 1 hour AND Restore the system operable status within 14 days or generate a Condition Report to document the event in accordance with SO-R-2.
			D	Not Applicable.
			E	Establish a continuous fire watch for those areas in which redundant systems or components could be damaged within 1 hour OR for those areas not having redundant systems or components establish roving fire watch patrol within 1 hour.

Table A-1. Comparison of Technical Specification Compensatory Actions to Current FPP Measures				
Affected System or Feature	Impairment	Original STS Action[18]	Current FPP Actions [19]	
			Plant	Action
			F	Establish a continuous fire watch with backup suppression equipment within 1 hour
			G	Establish a continuous fire watch with backup fire suppression capability for those areas containing redundant systems or components within 1 hour; For those areas not having redundant systems or components establish an hourly fire watch patrol within 1 hour. IF Required Action and associated Completion Times not met then initiate a Condition Report immediately.
Fire Hose Stations	One or more of the fire hose stations inoperable	Route additional equivalent capacity to fire hose to the unprotected areas from an OPERABLE hose station within 1 hour. STS Rev 2: route additional equivalent capacity within one hour if the inoperable fire hose is the primary mean of fire suppression; otherwise route additional hose within 24 hours. STS Rev 4: Changed wording to: With one or more of the fire hose stations inoperable, provide gated wye(s) on the nearest operable hose station. One outlet of the wye shall be connected to the standard length of hose provided at the station. The second outlet shall be connected to a length of hose sufficient to provide coverage for	A	Issue an Fire Protection System Impairment (FPSI) permit and within one hour: provide gated wye(s) on the nearest operable hose station(s) (one for the hose station and one to sufficient hose for the unprotected area); Post signs to indicate that the station is inoperable and which station is providing coverage; Post signs at the operable station indicating which hose is now providing coverage to the furthest area.
			B	Provide an alternate means of fire suppression for the unprotected areas within one hour, AND/OR Route and additional equivalent capacity fire hose to the unprotected area(s) from an OPERABLE hose station located in another fire zone using a gated wye off that operable station. Place signs at the backup fire suppression equipment to identify the proper hose to be used.
			C	Stage a hose of equivalent capacity which can service the unprotected areas from an operable hose station within one (1) hour from the time that the hose station is determined to be inoperable, if the inoperable hose station is the primary means of fire suppression; otherwise, stage the additional hose within 24 hours.

Table A-1. Comparison of Technical Specification Compensatory Actions to Current FPP Measures				
Affected System or Feature	Impairment	Original STS Action[18]	Current FPP Actions [19]	
			Plant	Action
		the area left unprotected. Where it can be demonstrated that the physical routing of the fire hose would result in a recognizable hazard to operating technicians, plant equipment, or the hose itself, the fire hose shall be stored in a roll at the outlet of the OPERABLE hose stations. Signs shall be mounted above the gated wye(s) to identify the proper hose to use. The above action shall be accomplished within 1 hour if the inoperable hose is the primary means of fire suppression; otherwise within 24 hours.	D	Route fire hose to provide equivalent nozzle flow capacity to the unprotected area(s) from the OPERABLE hose station or alternate water supply within 1 hour. (Required to run hose if operable water supply is not within 250' or area protected by the inoperable spray and/or sprinkler system)
			E	(ONE) If identified as being the primary means of fire suppression within one hour: provide gated wye(s) on the nearest operable hose station(s) (one for the hose station and one to sufficient hose for the unprotected area); Post signs to indicate that the station is inoperable and which station is providing coverage; Post signs at the operable and inoperable stations indicating which hose is now providing coverage to the furthest area; If not the primary means then actions to be completed within 24 hours. (ALL) Immediately notify the affected unit CRS/Shift Manager to suspend all hot work within the area affected until adequate compensatory measures have been established.
			F	Issue an Fire Protection System Impairment (FPSI) permit and within one hour: provide gated wye(s) on the nearest operable hose station(s) (one for the hose station and one to sufficient hose for the unprotected area); Post signs to indicate that the station is inoperable and which station is providing coverage; Post signs at the operable station indicating which hose is now providing coverage to the furthest area.

Table A-1. Comparison of Technical Specification Compensatory Actions to Current FPP Measures				
Affected System or Feature	Impairment	Original STS Action[18]	Current FPP Actions [19]	
			Plant	Action
			G	Provide an alternate means of fire suppression for the unprotected areas within one hour, AND/OR Route and additional equivalent capacity fire hose to the unprotected area(s) from an OPERABLE hose station located in another fire zone using a gated wye off that operable station. Place signs at the backup fire suppression equipment to identify the proper hose to be used.
STS Rev 2: Added the Section: *Yard Fire Hydrants and Hydrant Hose Houses*	One or more of the yard fire hydrants or associated hydrant hose houses are inoperable and are the primary means of fire suppression	Route sufficient additional lengths of 2.5" diameter hose located in an adjacent operable hydrant hose house to provide service to unprotected area(s) within 1 hours if it is the primary means of fire suppression; otherwise within 24 hours	A	Within 24 hr. attach sufficient additional lengths of 2.5 in. diameter hose located in an adjacent operable hydrant hose house to provide service to the unprotected area(s).
			B	NA - Two fire carts provided in lieu of hydrant hose houses, and contain fire fighting equipment necessary to support the fire brigade in response to a fire.
			C	Stage a hose of equivalent capacity which can service the unprotected areas from an operable hose station within one (1) hour from the time that a hose station is determined to be inoperable if the inoperable fire hose station is the primary means of fire suppression; otherwise, stage the additional hose within 24 hours.
			D	Establish backup suppression or verify available backup suppression on fire apparatus.
			E	Within 24 hours, verify sufficient additional lengths of adequate hose is available on an emergency response vehicle to provide service to the Protected Area.
			F	Provide sufficient additional lengths of 2 1/2" fire hose to provide service to the affected structure (Hydrant / Structures identified in the procedure)
			G	Stage adequate fire hose lines at the nearest operable hose station to ensure equivalent capacity backup hose protection to the unprotected area

Table A-1. Comparison of Technical Specification Compensatory Actions to Current FPP Measures				
Affected System or Feature	Impairment	Original STS Action[18]	Current FPP Actions [19]	
			Plant	Action
Fire Rated Assemblies (Fire Barrier)	One or more of the required barriers non-functional	With one or more of the required barriers non-functional, establish a continuous fire watch on at least one side of the affected penetration within 1 hour. STS Rev 2 added *or verify the operability of fire detectors on at least one side of the non-functional fire barrier and establish an hourly fire watch patrol.*	A	Hourly fire watch patrol –IF detectors or auto suppression on one side of the non-functional fire barrier – Otherwise Continuous Fire Watch
			B	Hourly fire watch patrol –IF detectors or auto suppression on one side of the non-functional fire barrier – Otherwise Continuous Fire Watch
			C	Hourly fire watch patrol –IF detectors or auto suppression on one side of the non-functional fire barrier – Otherwise Continuous Fire Watch
			D	Hourly fire watch patrol –IF detectors or auto suppression on one side of the non-functional fire barrier – Otherwise Continuous Fire Watch
			E	Hourly fire watch patrol –IF detectors or auto suppression on one side of the non-functional fire barrier – Otherwise Continuous Fire Watch
			F	Hourly fire watch patrol –IF detectors or auto suppression on one side of the non-functional fire barrier – Otherwise Continuous Fire Watch
			G	If Detection is operable in the room where the barrier is located establish Hourly Fire Watch, otherwise a Continuous Fire Watch is required

APPENDIX B: ADVANCED TECHNOLOGIES

In recent years, significant technological advances in fire technology have emerged which warrant consideration as possible alternatives to compensatory measures identified in the approved FPP. Depending on the plant-specific circumstances, these technologies may be found to provide an adequate level of compensation when used alone or in conjunction with the traditional measures specified in the approved FPP.

The following examples are only provided to illustrate advanced detection and suppression technologies that are currently available. Inclusion in this Appendix does not constitute an endorsement by NRC. In addition, depending on plant-specific circumstances (e.g., license basis requirements, fire hazards, physical construction, operating practices, etc.), the technologies may not be cost-effective, may be difficult to implement or may not provide an appropriate method of compensation.

B.1   Advanced Detection Technologies

Early detection is a key factor in preventing fire damage. The earlier a fire is detected, the sooner it can be controlled and extinguished. The initiation of combustion requires the conversion of fuel into a gaseous state by heating. The chemical decomposition of a solid substance by heat is called pyrolysis. During this phase, microscopic combustion particles are emitted that are often too small to actuate conventional ionization and photoelectric smoke detectors. While certain standard photoelectric and ionization detectors may be set to very high sensitivities, increasing sensitivity beyond the normal range typically results in a high number of nuisance alarms.

Recent advances in sensor types and signal processing technologies have led to the development of fire detection systems that are capable of detecting particles emitted during the pyrolysis or "incipient" [20] stage of fire growth. As shown in Figure B-1 below, detector actuation during the incipient stage may prevent fire damage by allowing early intervention; frequently before suppression is necessary.

---

[20] The National Fire Protection Association (NFPA) defines an incipient stage fire is a fire that is in the initial or beginning stage and which can be controlled or extinguished by portable fire extinguishers or small hose systems without the need for protective clothing or breathing apparatus. (Ref. 25)

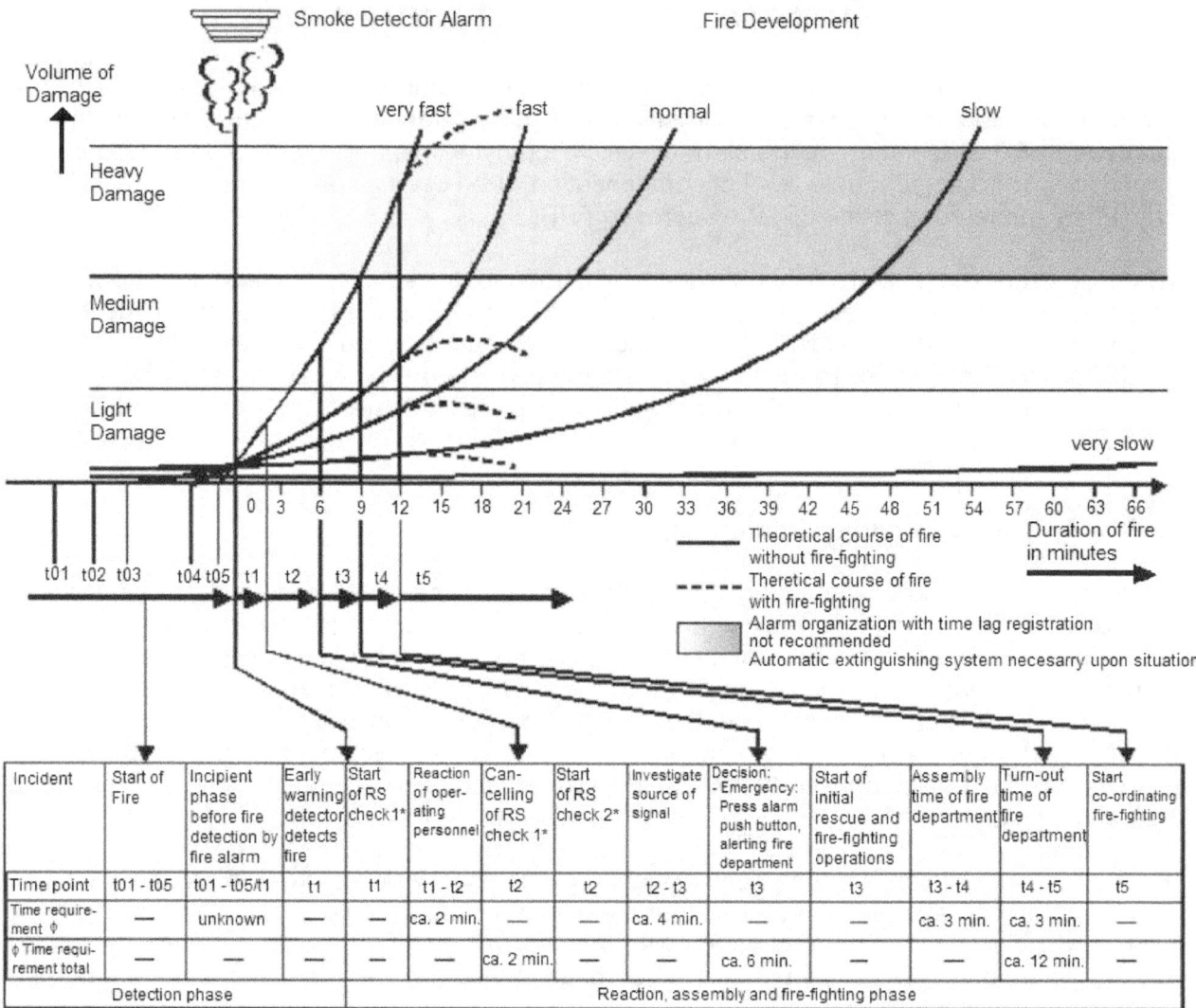

Figure B-1. Qualitative relationship between time and damage for different speeds of fire development and average detection, reaction and fire-fighting times (Ref. NUREG/CR-2409)

Systems that are able to detect products of combustion during pyrolysis are referred to as Very Early Warning Fire Detection (VEWFD) Systems. For a fire detection system to be considered a VEWFD system, it must meet the following sensitivity criteria specified in NFPA 76, "Standard for the Fire Protection of Telecommunication Facilities":

1. It must be set up to provide Alert thresholds of at least 0.2 percent per foot obscuration (effective sensitivity at each port), and,
2. Alarm thresholds of at least 1 percent per foot of obscuration (effective sensitivity at each port).

Detection technologies capable of satisfying the criteria in NFPA 76 for classification as a VEWFD system are available, and have been used extensively in the telecommunications industry for several years to guard against service disruptions caused by fire. It should be noted, that conventional spot type detectors installed to meet the requirements of NFPA 72 may be described as being able to detect incipient fires. However, to obtain the credit as a VEWFD system, the detection system must be capable of meeting the more stringent requirements in NFPA 76. Currently, the two primary types of VEWFD systems available are aspirated (air sampling) smoke detectors, and laser-based, spot-type, smoke detectors.

B.1.1   Aspirating (Air Sampling) Smoke Detectors

Thermal decomposition (pyrolysis) takes place causing solid materials to generate microscopic particles. Aspirating smoke detectors (ASDs) are capable of detecting these particles that evolve during pyrolysis which occurs well before the appearance of visible smoke. As discussed below, there are variations between specific designs. However, all ASD designs operate by actively and continuously sampling the air in a protected space.   An aspirator (aka fan) draws air through a series of small holes in a piping network to an optical measuring chamber. Inside the chamber a light source (e.g. high energy light source, laser or light-emitting diode [LED]) and optical sensor develop a signal that is proportional to concentration of smoke particles in the sampled air.

Obscuration, or the effect that smoke has on reducing visibility, is a unit of measurement which defines smoke detector sensitivity. A detector that requires higher concentrations of smoke to alarm will have a higher obscuration level (lower sensitivity). Systems used for very early warning of smoke and fire must be capable of sensing extremely low smoke levels, with obscuration values of 0.03% per foot or less. As shown in Table B-2 below, ASDs are capable of providing a much earlier warning of an impending fire than the traditional, spot-type ionization and photoelectric detectors.

Table B-2.   Detector Sensitivity Comparison (Ref. Brazzell, 2009)

Detector Type	Sensitivity (Obscuration per Meter)
Ionization	2.6–5.0% obs/m (0.8-1.6 % obs/ft)
Photoelectric	6.5–13.0% obs/m (2.0-4.2 % obs/ft)
Aspiration (ASD)	0.005–20.5% obs/m (.002-6.8 % obs/ft)

Over the past decade or so, several plants have installed aspirating smoke detectors in various plant locations. In addition, the use of aspirating smoke detectors has been found to provide a suitable approach under 10 CFR 50.48 (c) for assisting in the prevention of multiple spurious actuations that may occur as a result of fire within electrical cabinets and preventing fires within the cabinets from progressing to the point where a hot gas layer (HGL) would be developed in the room. (Ref.: Shearon Harris Nuclear Power Plant, Unit 1 – Issuance Of Amendment Regarding Adoption Of National Fire Protection Association Standard 805, "Performance-Based Standard For Fire Protection For Light Water Reactor Electric Generating Plants," ML101750602). Guidance on modeling the use of incipient fire detection systems in fire probabilistic risk assessment (PRA) applications at plants transitioning to a risk-informed approach under 10 CFR 50.48 (c) has been provided by the staff in its response to Frequently Asked Question (FAQ) 08-0046 (ML093220426).

Currently, there are three basic ASD sensor technologies: Cloud Chamber; Light Scattering and Laser Particle Counter. As illustrated in the following paragraphs, all aspirating smoke detectors use an aspirator (fan) to draw a sample of air from a series of pipes or tubes into a detection chamber. What differs is the technology used to measure the amount of smoke in the sampled air.

B.1.1.1 Cloud Chamber ASD

As shown in Figure B-2, in a cloud chamber smoke detector, a sample of air is drawn into a high humidity chamber. After the air sample is raised to a high humidity, the pressure is slightly reduced. If smoke particles are in the air, the moisture in the air condenses on them, forming a cloud in the chamber. The light obscuration of the cloud is a measure of the number of particles in the air sample (number density). The detector responds when the inferred number density is greater than a preset level. The cloud chamber system typically uses a valve and switching arrangement to sample from several detection zones.

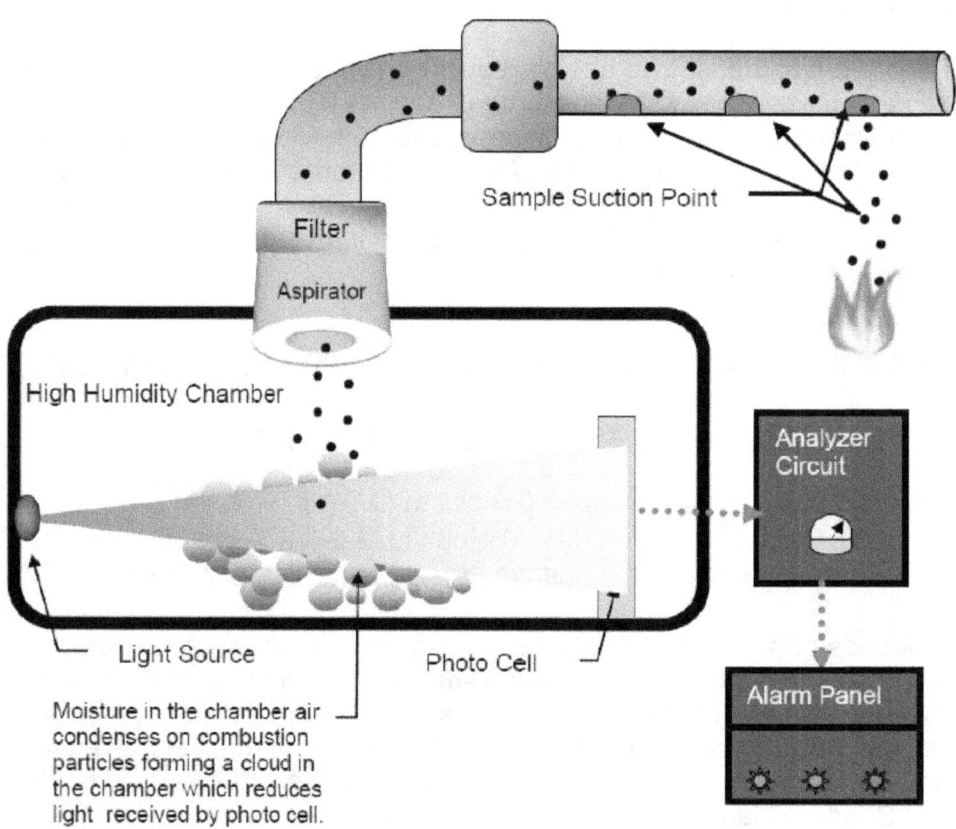

Figure B-2. Cloud Chamber

B.1.1.2 Light Scattering ASD

A stream of sampled air is continually passed through a detection chamber in which a high-energy light source is pulsed. Unlike the cloud chamber which senses light obscuration, the photosensitive device is not in the light beam path. As smoke particles enter the sensing chamber, light is reflected (scattered) from the smoke particles onto the photosensitive device causing the detector to respond. As illustrated in Figure B-3, an arrangement of light emitters, screening disk and receivers inside the measuring chamber keep emitted light signals from hitting the optical receiver cell during normal (no fire) conditions. Should smoke enter inside the box through the inlet apertures the floating smoke particles will scatter the light signal. Those scattered light rays will hit the optical cell and be transformed into an electric signal. The intensity of scattered light at a particular angular range is measured by a solid-state light receiver, and is proportional to the smoke concentration for a fixed smoke particle size distribution. The analyzer circuit triggers an alarm if a threshold value is exceeded for a predetermined number of consecutive pulses.

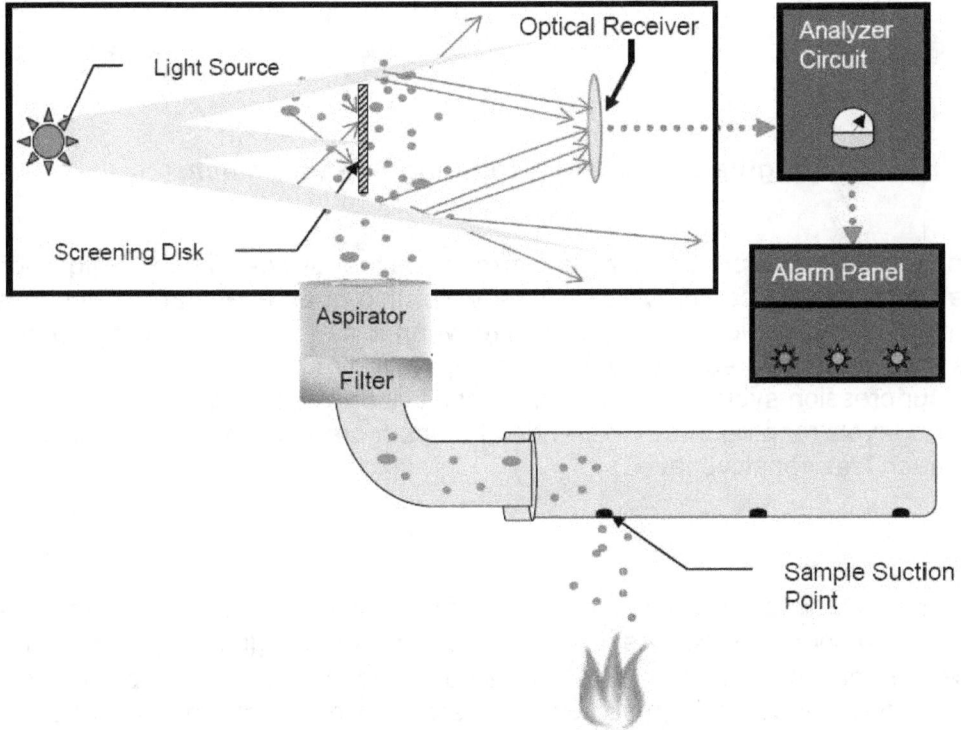

Figure B-3. Light Scatter Aspirating Smoke Detector

B.1.1.3 Laser Particle Counter ASD

Similar to the Cloud Chamber and Light Scattering technologies air is drawn into a detection chamber through a network of sampling pipes by an aspirator (Fan). Inside the chamber the laser particle ASD count the number of "flashes" caused by microscopic particles emitted but with much greater sensitivity. This measurement is typically used in conjunction with light scattering measurements to improve false alarm immunity.

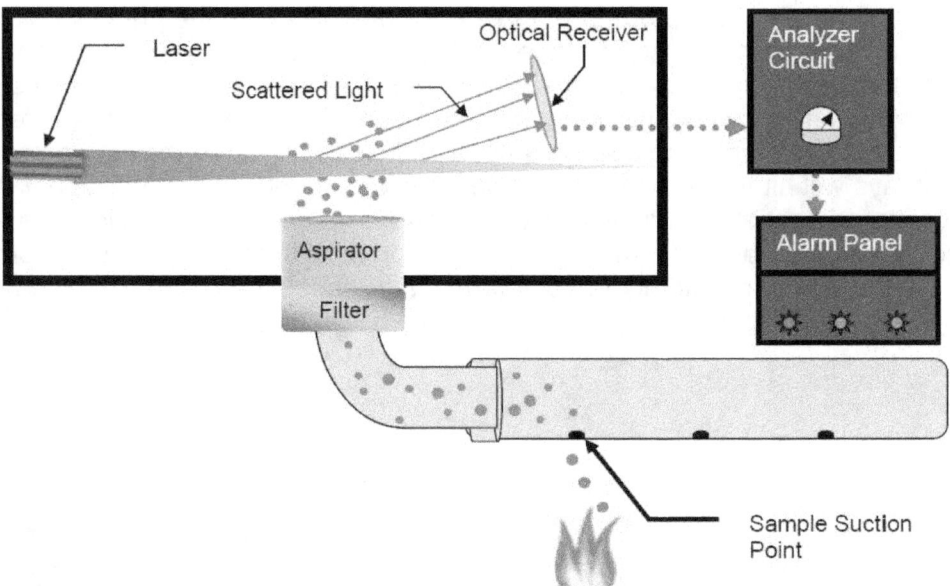

Figure B-4. Laser Aspirating Smoke Detector

ASD systems typically incorporate multiple alarm levels that are generally configurable. Therefore, an ASD system can provide very early warning of an event, prompting investigation at the earliest pyrolysis/smoldering stage of a fire when it is more easily extinguished. Other alarm levels may be configured to provide alarm inputs to fire protection systems, as well as activation of suppression systems. ASD alarm sensitivities are configurable and can be programmed to levels ranging from thousands of times more sensitive than a conventional detector, to much less sensitive level.

<u>Key Benefits</u>

- They are capable of providing indications of the earliest pyrolysis/smoldering stage of a fire allowing corrective actions to be taken before an extinguishing action is needed,
- Since the detector can be physically located outside the area to be protected, the system can be used in areas of high air flows, or harsh environments (e.g. vibration, humidity and temperature),
- They can provide early warning, irrespective of the level of stratification (air sampling ports can be sited at various horizontal and vertical positions),
- The ability to keep monitoring smoke levels after they set off an alarm may provide valuable data to emergency personnel, and,
- They afford ease of access for maintenance (compared to ceiling mounted detectors)
- Could contribute to the effort to maintain exposures to ionizing radiation "as low as is reasonably achievable" (ALARA) in high radiation areas.

Potential Limitations

- ASD may not be suitable for use in areas where a large fire may rapidly occur, such as those containing high concentrations of volatile combustibles, or in high voltage cabinets susceptible to energetic (arcing) faults,
- Sensor pipe flow degradations / obstructions may occur and could be difficult to locate,
- Dust may be a problem if the filtration of sample air is not adequate.

B.1.2   Laser-Based Spot-Type Smoke Detectors

Spot-type detectors, similar in appearance to traditional ceiling mounted detectors, are available that meet NFPA 76 criteria for application as a VEWDS. The principles of laser detection are similar to those of light-scattering photoelectric technology. In a photoelectric smoke detector, an LED emits light into a sensing chamber that is designed to keep out ambient light while allowing smoke to enter. Any particles of smoke (or dust) entering the chamber will scatter the light and trigger the photodiode sensor. Rather than a light-emitting diode (LED), laser spot detectors use a laser diode coupled to a lens, such that it creates a narrow but very intense light beam which provides up to 100x greater sensitivity than a standard photoelectric detector. If a particle of smoke enters the chamber, light from the laser is scattered and sensed by a photo detector. Algorithms built into the detector check the nature of the scattered light to determine whether the source is dust or smoke. If a determination of smoke is made, the alarm is signaled.

Laser-based detectors also provide multiple alarm set-points that generally are configurable In one design, users can select from nine different sensitivities in the range of 0.02–2% per foot obscuration for either pre-alarm or alarm settings.

Key Benefits

- Laser-based detectors can be mixed with standard photoelectric or ionization detectors on the same loop or system,
- An addressable system allows identification of the exact location of fire,
- They can detect both fast-flaming and slow-smoldering fires.

Potential Limitations

- Laser-based detectors may not be suitable for use in areas containing large concentrations of dust or areas because they potentially might give false alarms in areas where welding or other processes are undertaken that generate combustion particles.

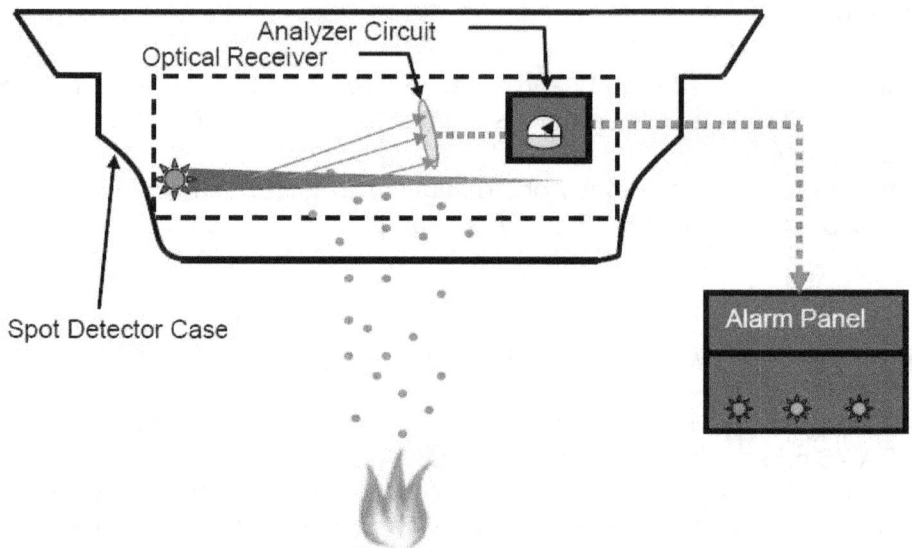

Figure B-5. Spot-Type Laser Detector

B.1.3   Video Image Detection

Recent improvements in video camera capabilities, computer processing, and image analysis, combined with a desire for real time monitoring capabilities have driven the development of advanced video image detection (VID) systems similar to the one illustrated in Figure B-6. Advanced VID systems should not be confused with standard closed-circuit television (CCTV) systems such as those used for plant security which only provide an image. While VID systems may use the same cameras, in a VID system, the video image from the camera is processed by proprietary software to determine if smoke or flame from a fire can be identified in the video image. The detection algorithms identify the flame and smoke characteristics based on spectral, spatial or time-based properties such as changes in brightness, contrast, edge content, motion, dynamic frequencies, and pattern and color matching. Unlike conventional fire detectors, VID is not governed by a single physical principle, such as temperature or optical obscuration. Instead, several software algorithms detect features in the video that correspond to one or more visible characteristics of fire.

The National Fire Alarm and Signaling Code, NFPA 72, covers the application, installation, location, performance and maintenance of fire alarm systems and their components. The use of VID systems for flame and smoke detection was first recognized in the 2007 edition of the National Fire Alarm Code[21].

Multiple environmental and system variables must be considered in designing and using a VID system, including obstructions, changing light levels, ventilation, lens contamination, and camera settings. Depending on the particulars of an application, the installation goals and performance criteria could vary significantly. Consequently, NFPA 72 specifies a performance-

---

[21] As part of the 2010 revision, the title was changed from "National Fire Alarm Code" to "National Fire Alarm and Signaling Code."

based design approach in the form of an engineering evaluation. This means that VID systems should be designed to achieve a specified goal for a specified use or application.

Key Benefits

- VID systems have the ability to protect a large area, while achieving fast detection. They can detect smoke or flame anywhere within the field of view of the camera, whereas conventional smoke detectors require smoke to migrate to the detector.
- They can be used for outdoor applications.
- Digitized video and/or audio streams can be sent to any location via a wired and/or wireless IP network, enabling video monitoring and recording from anywhere with network access.
- They can use the system's basic hardware (i.e., the cameras and wiring) for multiple purposes (e.g., fire, flood, security, equipment monitoring).
- They improve the operator's response and positive event verification.
- Video archiving of events supports future investigations of fire.
- They can operate in environments where spot detectors may not be effective.
- Instant situational awareness reduces the operator's and fire brigades' response time.
- Could contribute to the effort to maintain exposures to ionizing radiation "as low as reasonably achievable possible" (ALARA) in high radiation areas.

Potential Limitations

- VID systems require a certain minimum amount of light for effective detection and most will not work in the dark.
- Depending on the software algorithms, nuisance alarms may be generated by events other than fires, such as steam from vents or exhaust from vehicles. It should be noted that the potential for nuisance sources is highly dependent on the specific VID technology. Some systems have the abilities to ignore areas of the field of view that may have potential nuisance sources, to adjust sensitivity, and to adjust the persistence time of the event before an alarm signal is issued. Manufacturers also have developed specific alarm algorithms to avoid signaling common nuisance events.
- VID systems should not be expected to perform in the environments outside the normal operating space of general CCTV installations (e.g., pointing a camera where the sun can be in field of view).
- Video images may be degraded by environmental contaminants or hardware adjustments that change the image's focus and brightness. Self-diagnostic capability typically is required to determine video image quality for proper detection performance.
- Because VID is a line-of-sight device, it typically requires an unobstructed view of the area to be protected.

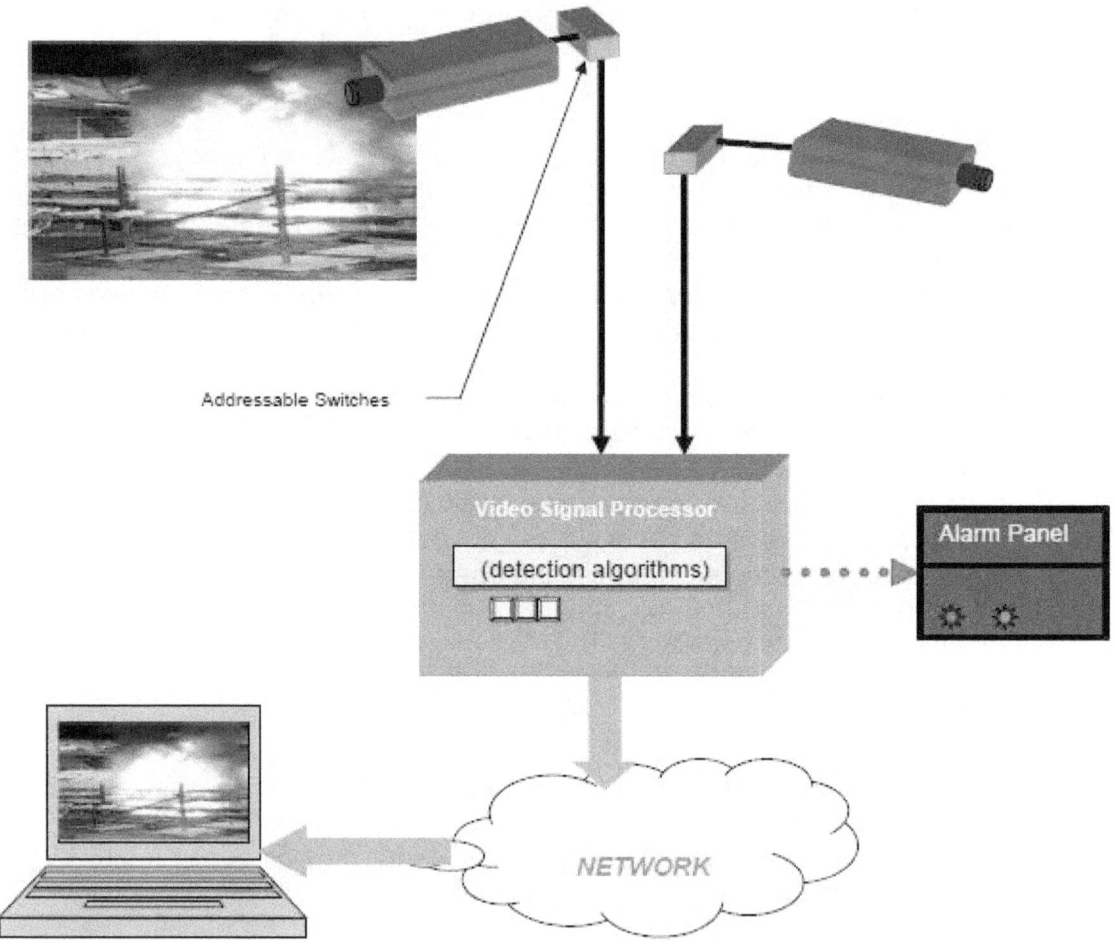

Figure B-6. Video Image Detection

B.1.4   Thermal Imaging Cameras

One tool currently available to identify possible fire initiators is the thermal imaging camera. The advantage of this device is its ability to rapidly display areas of varying temperatures without contact. This allows for the identification of specific equipment elements that are operating at higher temperatures than adjacent elements.

Thermal imaging instrument measures radiated infrared (IR) energy and convert the data to corresponding maps of temperatures. With thermal imaging technology, cameras capture this energy. Once captured, the camera's processor software reads the infrared signals and translates them into an image that we can easily see, providing an early warning of hot spots that are detected. Because thermal imaging cameras use the heat emitted from an object to produce images, they are able to work in the dark. Using this technology, it is possible to "see" the surface temperature of any object. A true thermal image is a gray scale image with the hot items shown in white and the cold items in black. Temperatures between the two extremes are shown as gradients of gray. Some thermal imagers enhance operator performance by adding color which is artificially generated by the camera's electronics in response to the thermal attributes seen by the camera.

Infrared (IR) spot measurements also are obtainable using a simple IR point radiometer (sensor) that measures the object's emitted IR energy and converts it into a digital temperature readout. However, a point radiometer does not provide an image of the object and, thus, it is difficult to find what or where the problem is precisely located without actually scanning the entire object or surface.

Thermography may serve as an alternative or complementary approach to conventional fire detection technologies provided the system includes some basic features such as the ability to

- detect and clearly visualize emerging hot spots
- measure and indicate temperature
- raise an alarm when a temperature threshold is exceeded.

Often, deterioration, such as high contact resistance in a circuit breaker contact, or loose electrical connections produce warning signs that can be identified by a thermal imaging device before the device fails. Using a hand-held infrared camera, potential precursors to fire ignition can be quickly identified for evaluation. In many cases, precursors can be detected well before an actual failure. Many NPP fire brigades currently use this technology as hand-held thermal imaging cameras (TICs) to aid in locating and evaluating fire conditions. Other applications include locating potential ignition sources prior to fire, such as loose electrical connections, failing/overheated transformers, and, overloaded motors or pumps.

Advantages

- These cameras allow monitoring of locations that are difficult to reach due to extreme environmental conditions.
- They provide early identification of potential ignition sources, such as hot spots in electrical switchgear enclosures or transformers.
- They can significantly enhance the capabilities of a roving fire watch and may justify less frequent tours of certain areas.
- They offer a warning before an actual fire event.
- Could contribute to the effort to maintain exposures to ionizing radiation "as low as is reasonably achievable" (ALARA) in high radiation areas.

Potential Challenges

- Conditions that cause the camera's thermal detector to become saturated, or in which the range of temperatures detected becomes too wide for the optics and/or electronics to operate at the highest resolution.
- Thermal images may be difficult to interpret. Reflections and low surface emissivity can produce false indications – training is required.

B.2   Advanced Suppression Technologies

Research into the development of Halon alternatives has resulted new fire extinguishing agents which have been incorporated into the design of completely self-contained fire suppression systems and devices. The systems typically use clean agents that are non-conductive and leave no residue. Clean Agents are particularly useful for hazards where:

- An electrically non-conductive agent is required
- Cleanup of other agents presents a problem
- Hazard obstructions require the use of a gaseous agent
- The hazard is normally occupied and requires a non-toxic agent

Types of hazards typically protected with clean agents include:

- Computer rooms
- Control rooms
- Telecommunications facilities
- Electric switchgear

Although clean agents are most common, the systems are generally compatible with most commercially available fire suppression agents. The suppression of fires by clean agents is covered by the NFPA 2001, Standard on Clean Agent Fire Suppression Systems. . The self-contained design of the examples illustrated in the following paragraphs enables them to provide a standalone, transportable, fire suppression system.. Depending on the plant-specific circumstances, they may provide an effective alternative to compensatory measures specified in the FPP. .

B.2.1   Solid Propellant Gas Generators

Based on automotive airbag technology, gas generators have been developed for fire suppression applications. Gas generators can produce a large quantity of gases (mainly N2, CO2 and water vapor) by combustion of solid propellants.
. Solid propellant gas generators typically consist of solid propellant tablets which rapidly generate a gas when ignited. Two common types of solid propellant gas generators are condensed aerosol extinguisher which produces a powdered aerosol and a nitrogen generator that produces inert nitrogen gas.

B.2.1.1   Condensed Aerosol Generators
In a condensed system the aerosols are produced by exothermic chemical reaction (i.e., pyrotechnically) using a solid compound. When the system is activated the aerosol is introduced into a space through a delivery system similar to that used for gaseous agents. Condensed aerosol generators are completely self-contained and suitable for nearly all types of plant fire hazards.

The condensed aerosol generator can be activated manually, thermally (by the fire) or electrically from a suitable detection device. Different-sized generators are available to protect different volumes. In addition, several generators can be strung together to provide additional coverage. The design, installation, operation, testing, and maintenance of condensed aerosol fire-extinguishing systems are governed by NFPA 2010, Standard for Fixed Aerosol Fire-Extinguishing Systems.

As illustrated in Figure B-7 a condensed aerosol generator is completely self-contained within a heavy duty aluminum canister that can be easily installed as needed. Storing the extinguishing agent in a dense solid form until activated, permits the condensed aerosol generators to be fairly small. As a result, the canisters may be hung from walls, ceilings or mounted within an enclosure.

Key Benefits

- Compact
- Lightweight
- No pipework required
- Environmentally friendly
- Ease of installation
- Long service life
- High extinguishing efficiency

Possible Limitations

- Reduced visibility in protected space following discharge
- The temperature of the expelled aerosol near the generator surface may be as high as 200 °C (392°F).
- Are effective only in closed spaces. May not be suitable for ventilated enclosures
- Should not be used in normally manned enclosures.

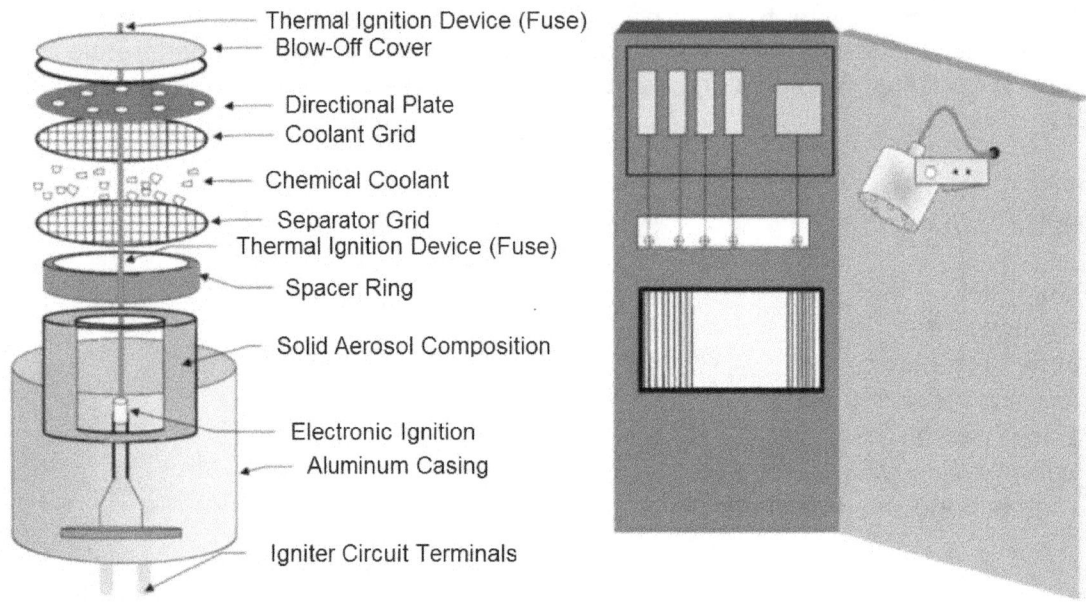

Thermal Ignition Device (Fuse)
Blow-Off Cover

Directional Plate
Coolant Grid

Chemical Coolant

Separator Grid
Thermal Ignition Device (Fuse)

Spacer Ring

Solid Aerosol Composition

Electronic Ignition
Aluminum Casing

Igniter Circuit Terminals

Figure B-7. Condensed Aerosol Generator

B.2.1.2   Nitrogen Gas Generators

As shown in Figure B-8, Nitrogen (N2) generators are physically similar to condensed aerosol
generators. Like the condensed aerosol generator, the N2 generator does not need high-
pressure gas storage cylinders and associated piping networks. Nitrogen generators produce
an inert N2 gas, which reduces the concentration of oxygen in a room below the level that will
sustain combustion. However, the oxygen concentration is maintained at a sufficient level to
meet the requirements of NFPA 2001 for clean agent Halon 1301 alternatives in normally
occupied areas.

N2 gas generation is initiated in response to an electrical signal that is generated by a fire
detector, fire control panel or manual pull station. The electrical signal actuates a pyrotechnic
device (squib) which ignites the solid N2 fuel. Suppression coverage varies with the size of the
generator canister. One manufacturer states that a single 6" x 12" generator will cover up to
200 cubic feet. Individual generators can be installed under floors, on ceilings or inside cabinets
and multiple generators can be daisy-chained together to provide a total room flooding
capability.   N2 systems should conform to NFPA 2010 Standard On Aerosol Fire-Extinguishing
Systems - Edition 1 and UL 2775 - Fixed Condensed Aerosol Extinguishing System Units -
Edition 1.

Key Benefits

- Long (up to 25 year) service-free shelf life
- Rechargeable on site

- Compact
- No piping or nozzles
- Releases harmless, inert, N2 gas - maintains oxygen levels that are safe for occupied spaces
- No residue or cleanup

Potential Limitations

- enclosure must be capable of holding the gas and be able to withstand the pressure produced during discharge

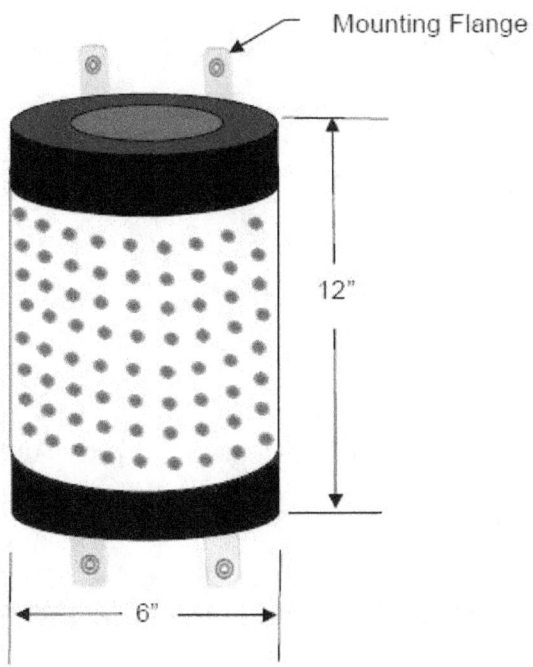

Figure B-8. N2 Gas Generator

B.2.2 Transportable Automatic Suppression Systems

Several self-contained, transportable, fire suppression systems are commercially available that are capable of detecting and extinguishing fire inside equipment, cabinets or enclosed spaces r up to 250 cu ft. As illustrated in Figure B-9, these systems incorporate flexible, polymer tubing that acts as both a fire detector and suppressant dispersion nozzle. When exposed to the heat of fire (typically 100°C) the tubing ruptures to dispense the suppression agent. A key advantage of these systems is that the flexible polymer tubing can be routed to provide protection in many different types of enclosures, including cable trays. In addition to being highly transportable these systems do not require any type of outside electrical source or detection system and will remain totally operational during power outages.

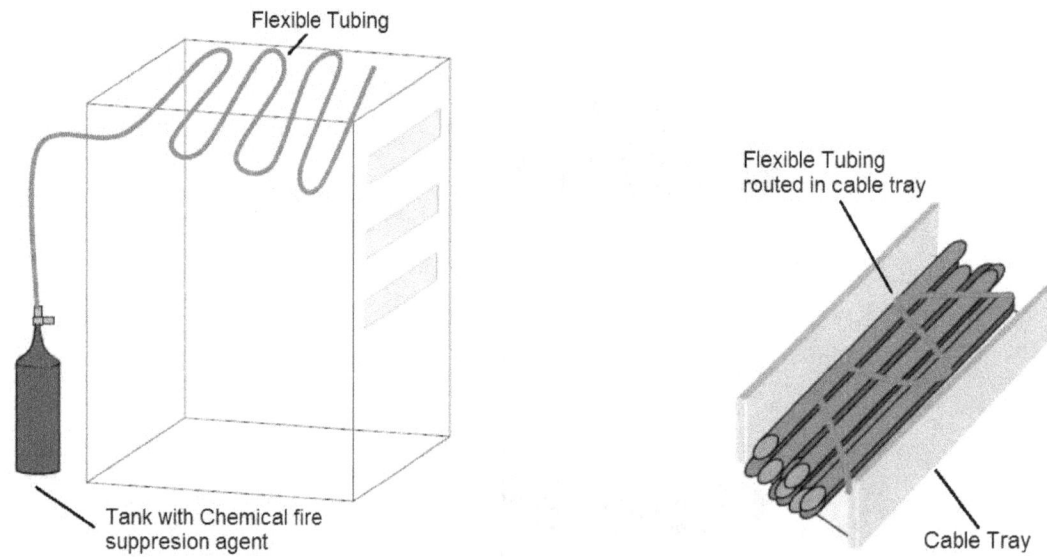

Figure B-9. Transportable Suppression System Applications

B.2.3 Pre-packaged Portable Water Mist Systems

Water has several favorable properties for fire suppression. When applied in a fine mist, its effectiveness is further improved by the increased surface area of water that is available for heat absorption and evaporation. Water mist systems have been demonstrated to extinguish a wide variety of fires, including fires in electrical and electronic equipment cabinets. In addition, evacuation of the compartment may not be necessary and the electronic equipment can continuously be operated during the discharge of a water mist especially if a zoned system is used (Ref. Liu, 2001).

NFPA 750, Standard on Water Mist Fire Protection Systems, contains the minimum requirements for the design, installation, testing, and maintenance of such systems. Pre-packaged, self-contained, portable water mist systems are of particular interest for their possible use as an alternate compensatory measure. As illustrated in Figure B-10, a completely self-contained system is commercially available that is pre-packaged on steel plate skid. The water cylinder is bolted to the skid and the nitrogen cylinder(s) are mounted to the water cylinder with straps. The skid is designed so it can be easily transported with a fork lift. According to one manufacturer, this system has successfully demonstrated fire extinguishment in Factory Mutual fire tests for machinery spaces[22] up to 9,175 ft³ (260 m³).

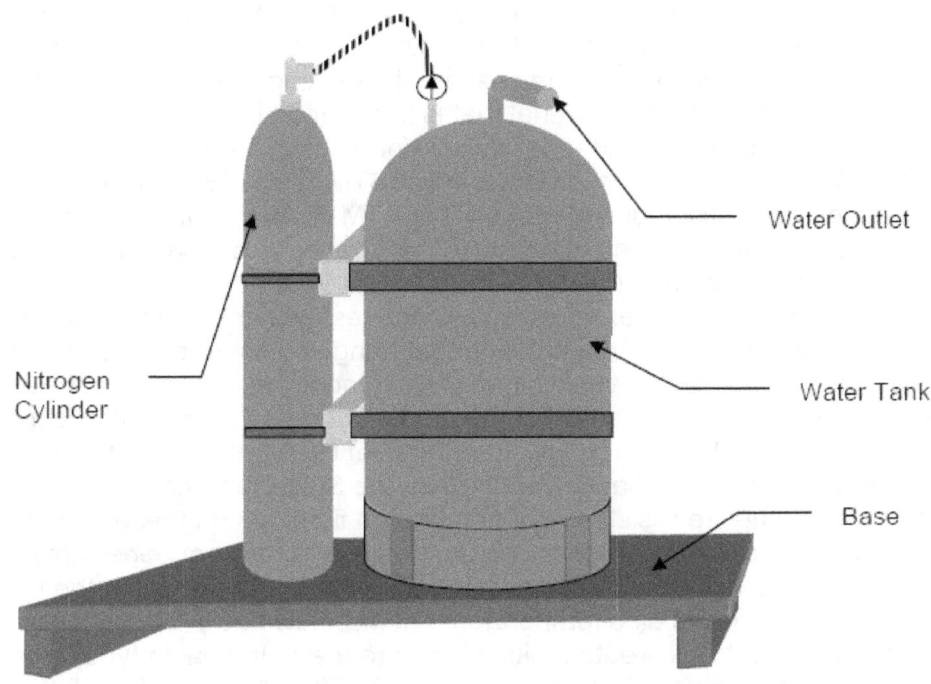

Figure B-10. Skid-mounted Water Mist System

B.3   Temporary Fire Barrier Penetration Seals

To limit fire damage and prevent the spread of toxic products of combustion (e.g. smoke, hot gases and fumes) NPPs are divided by fire-rated structural barriers into separate fire areas. The fire-rated walls, floors and ceiling assemblies (structural fire barriers) have sufficient fire resistance to withstand the fire hazards associated with the area and to protect important equipment located in the area from a fire that occurs outside the area. These structural fire barriers may contain a number of penetrations or openings to allow such services as piping, electrical conduits, cable trays; and ventilation ducts to pass from one fire area to another. To maintain the fire-resistive integrity of the barriers and provide reasonable assurance that a fire will be confined to the area in which it started, openings and voids in structural fire barriers are closed with penetration seal assemblies (also known as firestops) that have been tested and qualified to ensure the fire resistance rating of the barrier is maintained.

According to NUREG-1552, (Ref. US NRC, NUREG-1552), each NPP has an average of 3,000 penetration seals. However, a single NPP may have as many as 10,000. The fire at the Browns Ferry NPP in 1975 illustrated how improper penetration seal materials can contribute to the spread of fire. Today, penetration seal materials are available that, if properly installed and

---

[22] A machinery space is defined as areas that contain flammable liquid processing hazards with Class 1, 2, or 3 flammable liquids as specified in NFPA 325 and incidental Class A combustibles

maintained, will maintain the fire-resistive integrity of the fire barriers in which they are installed. A fire barrier penetration seal is a system that includes the fire-rated wall or floor, the opening, the cable or conduit that passes through the opening, and the material used to seal the opening. The performance of a penetration seal is dependent upon the specific assembly of materials tested including the number, type, and size of penetrations and the floors or walls in which it is installed. Achieving an effective penetration seal requires taking all of the components present into consideration when selecting the material that is best suited for that application.

Fire barrier penetration seals are intended for use in openings in fire-resistive walls, floors, and ceilings that have been evaluated in accordance with ASTM E119, "Standard Test Methods for Fire Tests of Building Construction and Materials."[23] ASTM E119 assigns ratings based on "T" (temperature rise) and "F" (time for flame spread). The T-rating is a measure of the thermal conductivity of a firestop system and can be considered a temperature rating. This is the time required for various points on the unexposed side of the test assembly to rise 325 degrees over the starting (ambient) temperature. As a result, penetrating items which conduct heat readily, such as metallic pipes and conduits, would significantly hinder the ability to achieve a acceptable T-rating when tested under ASTM E119. To address this and other concerns, ASTM E814-11a, " Standard Test Method for Fire Tests of Penetration Firestop Systems" and IEEE Std. 634, "IEEE Standard for Cable-Penetration Fire Stop Qualification Test" have been developed to test the unique fire resistance of penetration firestops. In these standards the T-rating is intended to provide engineering information relative to how hot penetrants might become in a fire exposure. A T-rating of a specified duration establishes that the fire stop effectively maintained unexposed side temperatures at or below 325°F above initial temperature for a general acceptance test or the auto ignition temperature of the cable type tested for a cable-specific qualification. Conformance to either ASTM E814-11 or IEEE Std. 634 a provides assurance that the penetration seal (firestop) meets its design objective of ensuring that the penetrated fire barrier will be capable of withstanding the fire hazards located within the area of concern and will adequately protect important equipment located in the area from a fire that occurs outside the area.

All fire barriers and fire barrier penetration seals do not have the same level of safety significance. The importance of a specific fire barrier depends on many factors, such as the importance of the equipment in the fire area (and adjacent areas); the configuration and location of combustible materials and other fire hazards, if any, in the areas; the potential for fire growth in the areas; the other fire protection features installed in the areas; and the accessibility of the areas to the plant fire brigade. Similarly, the importance of a specific fire barrier penetration seal depends on these factors and on such other factors as its size, location or position in the fire barrier, and the number and sizes of the other seals in the barrier.

In Generic Letter 86-10 (Ref. US NRC, GL88-10) the staff established that certain penetration seals need not have the same fire rating as the barrier in which they are installed and that certain fire barrier penetrations may not need to be sealed at all provided they are considered in the plant in its evaluation of the effectiveness of the overall barrier. Specifically, Interpretation 4, "Fire Area Boundaries," states, in part:

> The term "fire area" as used in Appendix R means an area sufficiently bounded to withstand the [fire] hazards associated with the area and, as necessary, to protect important equipment within the area from a fire outside the area. In order to meet the regulation, fire area boundaries need not be completely sealed floor-to-ceiling, wall-to-

---

[23] NRC classically has used NFPA 251 rather than E119 although they are the same.

wall boundaries. However, all unsealed openings should be identified and considered [in] evaluating the effectiveness of the overall barrier. Where fire area boundaries are not wall-to-wall, floor-to-ceiling boundaries with all penetrations sealed to the fire rating required of the boundaries, licensees must perform an evaluation to assess the adequacy of fire boundaries in their plants to determine if the boundaries will withstand all [fire] hazards associated with the area.

Thus, for penetration seals that cannot be demonstrated to meet ASTM E814-11a, an engineering evaluation may be used to determine the expected fire resistance rating. Licensees evaluate such seals on a case-by-case basis. The engineering evaluations performed to assess the effectiveness of the penetration seals are based on the expected fire resistive performance of the seal and on the fire hazards and fire protection features in the fire area. These analyses may use computer simulation and mathematical fire modeling, thermodynamics, heat-flow analysis, and materials science to predict the fire performance of the penetration seal assembly. Deviations from NRC requirements or accepted industry standards for fire barrier penetration seals should be technically substantiated as part of the review and approval of the fire protection plan or in other separate formal correspondence.

Like other plant features, it is expected that over the life of the plant, instances of degraded fire barrier penetration seals will be found. It is also expected that penetrations will be opened to allow for plant modifications, maintenance and upgrades. When such conditions are identified, NRC approved fire protection programs typically specify that fire watches be posted to maintain an appropriate level of defense in depth to assure the deficiency will not pose an undue risk to public health and safety. In general, if fire detection is not available on either side of the barrier, the FPP will specify the establishment of a continuous fire watch. Typically, in cases where automatic detection systems protect the affected components, an hourly fire watch patrol is specified. Although posting a fire watch is the most *common* compensatory measure, depending on the plant-specific conditions and nature of the deficiency, a number of products are currently available that may provide a more *effective* compensatory measure. Two specific examples are described below.

B.3.1   Intumescent Pillows and Blocks

An intumescent substance expands or swells when exposed to heat. This phenomenon is the working principle behind many fire barrier penetration seal products: they expand when exposed to the heat of fire thereby closing any small voids or gaps that may remain after installation or formed during a fire by melted components.

Intumescent pillows and blocks are two examples of temporary fire-rated materials that are available to seal medium to large openings in structural fire barriers. The pillows and blocks may be used to temporarily fill openings that are entirely blank or around penetrating items such as pipes, conduits, cable trays and HVAC ducts. This capability could be especially useful when a fire barrier is removed or breached during plant modifications. If a fire does not occur, they may be readily removed and stored for reuse. However, should a fire occur, the bags or blocks will expand, tightly closing any small spaces between cables, trays and masonry to create a fire-rated seal that is capable of withstanding mechanical damage caused by falling debris or a hose stream from fire fighters.

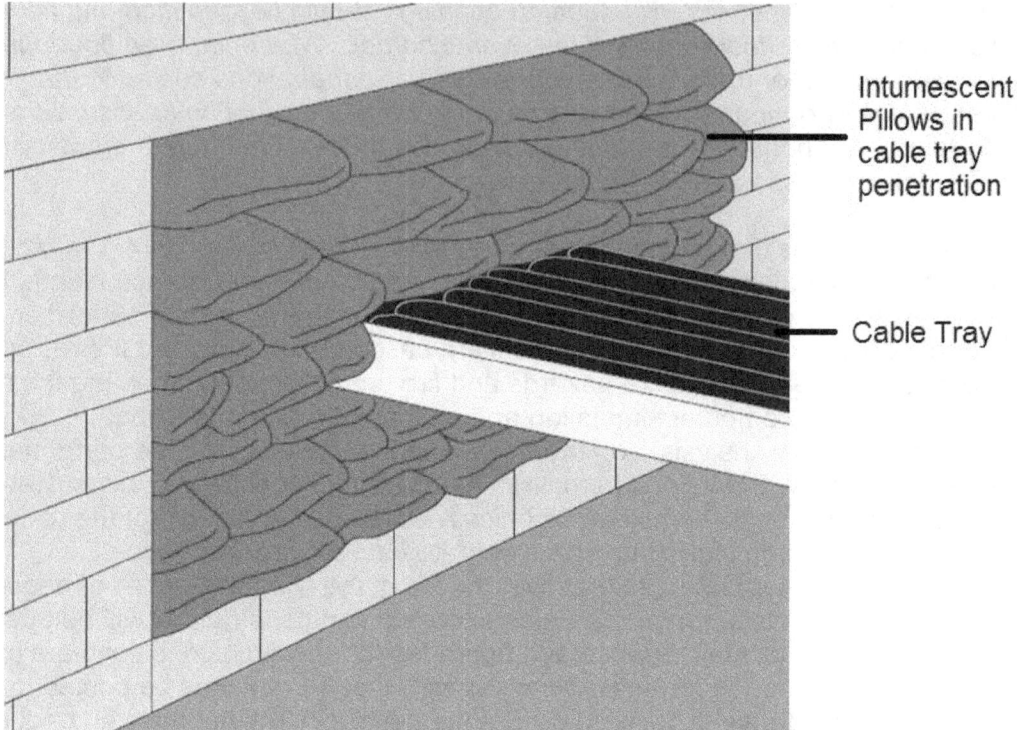

Figure B-11. Diagram of Intumescent Pillows in a cable tray penetration

Key Benefits

- Expand and seal into place when exposed to high temperatures
- Can provide up to 4-hours of fire resistance (F-rating) when tested in accordance with ASTM E 814
- Relatively easy to install, remove, and replace
- Useful for sealing difficult configurations such as where access is restricted to one side of the assembly
- Can be reused
- Pillows may be used to temporarily protect cable trays during cutting and welding
- Long shelf life

Potential Challenges

- May be subject to damage during use and un-intentional or unauthorized removal - periodic inspection is necessary
- Some configurations may require the installation of wire mesh.
- Excessive handling or abuse may permanently compress the pillows.

B.3.2   Reusable Firestop Plugs and Putty

For small to medium sized fire barrier penetrations and openings various plugs and putties are available which may provide an effective temporary seal. Empty conduit penetrations created during plant maintenance or cable routing modifications may be quickly sealed by firestop plugs. Plugs having a three-hour fire rating are available and are easily removed for penetration access. The plugs are typically prefabricated to fit standard conduit /pipe sleeve sizes. However, custom sizes are available.

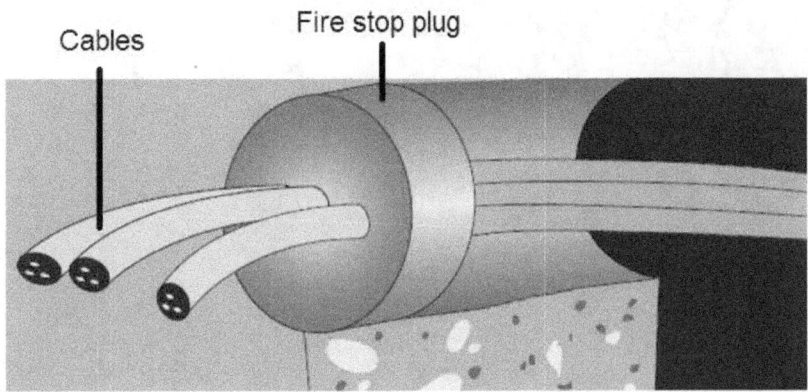

Figure B-12. Diagram of fire stop plug in a cable penetration

For sealing small diameter conduits or openings around metallic pipes, fire stop putty, may provide an effective temporary seal. Fire stop putty is a ready-to-use fire stop product that is hand pressed into place. During normal use the putty remains soft and pliable. However, in response to high temperatures caused by fire, the putty will expand within the opening, forming a solid char that helps prevent through penetration of the fire. Since no drying or curing is required, the putty is fully functional as soon as it is installed.

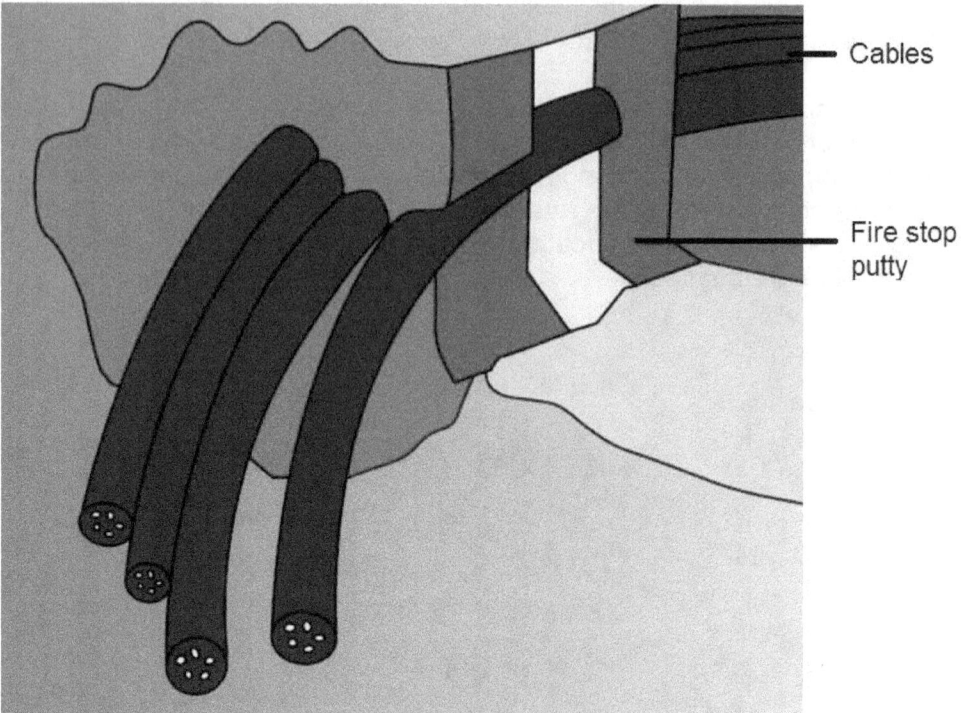

Figure B-13. Diagram of fire stop putty in a cable penetration

Key Benefits

- Can provide up to 3-hour F-ratings of fire resistance when tested in accordance with ASTM E 814
- Reusable
- Smoke and gas tight
- Weather resistant
- Long shelf life
- Quick and easy installation: no special tools required
- Does not require cable de-rating
- No curing or drying time: provides an immediate firestop
- Intumescent
- Easily re-penetrated

Potential Challenges

- May be subject to damage during use and un-intentional or unauthorized removal. Therefore, periodic inspection is necessary
- Putties contain oils which may be absorbed into porous surfaces. Sleeves are recommended
- Only high-tack putties should be chosen. Less tacky products may not stay in place

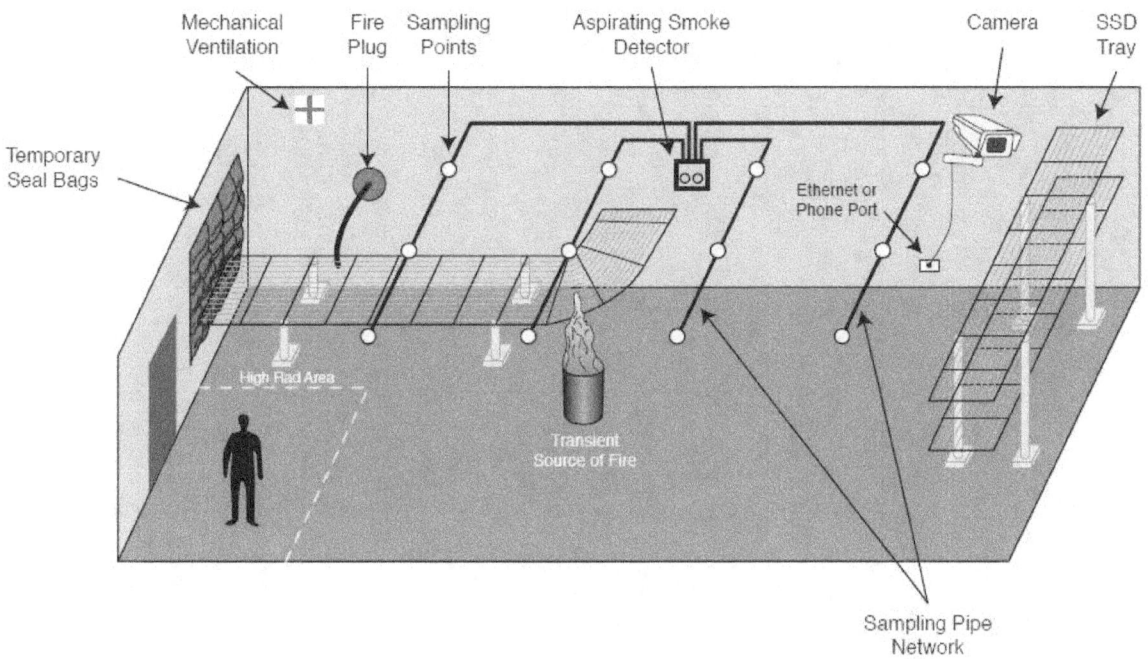

Figure B-14.   Possible Applications of Advanced Technologies

NRC FORM 335	U.S. NUCLEAR REGULATORY COMMISSION	1. REPORT NUMBER
(12-2010)		(Assigned by NRC, Add Vol., Supp., Rev.,
NRCMD 3.7		and Addendum Numbers, if any.)
	BIBLIOGRAPHIC DATA SHEET	
	(See instructions on the reverse)	NUREG/CR-7135
		Draft

2. TITLE AND SUBTITLE	3. DATE REPORT PUBLISHED	
Compensatory and Alternative Regulatory MEasures for Nuclear Power Plant FIRE Protection (CARMEN-Fire), Draft Report for Comment	MONTH	YEAR
	June	2013
	4. FIN OR GRANT NUMBER	
	N6760	

5. AUTHOR(S)	6. TYPE OF REPORT
K. Sullivan	Technical Report
	7. PERIOD COVERED (Inclusive Dates)
	Mar/22/1975 to 2009

8. PERFORMING ORGANIZATION - NAME AND ADDRESS (If NRC, provide Division, Office or Region, U. S. Nuclear Regulatory Commission, and mailing address; if contractor, provide name and mailing address.)

Department of Energy, Science and Technology
Brookhaven National Laboratory
P.O. Box 5000

9. SPONSORING ORGANIZATION - NAME AND ADDRESS (If NRC, type "Same as above", if contractor, provide NRC Division, Office or Region, U. S. Nuclear Regulatory Commission, and mailing address.)

Fire Research Branch
Division of Risk Analysis
Office of Nuclear Regulatory Research

10. SUPPLEMENTARY NOTES
NRC Project Manager: F. Gonzalez

11. ABSTRACT (200 words or less)
Employing compensatory measures, on a short-term basis, is an integral part of NRC-approved fire protection programs. However, compensatory measures are not expected to be in place for an extended period of time. The NRC staff expects that the corrective action(s) will be completed, and reliance on the compensatory measure eliminated, at the first available opportunity, typically the first refueling outage. Thus, a compensatory measure that is in place beyond the next refueling outage (typically 18 – 24 months) is considered to be a "long-term compensatory measure."

This report is intended to serve as a reference guide for agency staff responsible for evaluating the acceptability of interim compensatory measures provided to offset the degradation in fire safety caused by impaired fire protection features at nuclear power plants. The report documents the history of compensatory measures and details the regulatory framework established by NRC to ensure they are appropriately implemented and maintained. This report also explores technologies that did not exist when the current plants were licensed such as video-based detection, temporary penetration seals and portable suppression systems which under certain conditions may provide an effective alternative to traditional measures specified in a plant's approved fire protection program.

12. KEY WORDS/DESCRIPTORS (List words or phrases that will assist researchers in locating the report.)	13. AVAILABILITY STATEMENT
Compensatory Measures	unlimited
Fire Protection	14. SECURITY CLASSIFICATION
Alternative Compensatory Measures	(This Page)
	unclassified
	(This Report)
	unclassified
	15. NUMBER OF PAGES
	16. PRICE

NRC FORM 335 (12-2010)

UNITED STATES
NUCLEAR REGULATORY COMMISSION
WASHINGTON, DC 20555-0001

OFFICIAL BUSINESS

NUREG/CR-7135
Draft

Compensatory and Alternative Regulatory MEasures for
Nuclear Power Plant FIRE Protection (CARMEN-FIRE)

June 2013

www.ingramcontent.com/pod-product-compliance
Lightning Source LLC
Chambersburg PA
CBHW080302180526
45167CB00006B/2641